Bulgarian Studies in the Philosophy of Science

BOSTON STUDIES IN THE PHILOSOPHY OF SCIENCE

VOLUME 236

BULGARIAN STUDIES IN THE PHILOSOPHY OF SCIENCE

Edited by

DIMITRI GINEV
St. Kliment Ohridski University of Sofia, Bulgaria

KLUWER ACADEMIC PUBLISHERS
DORDRECHT / BOSTON / LONDON

A C.I.P. Catalogue record for this book is available from the Library of Congress.

ISBN 1-4020-1496-1

Published by Kluwer Academic Publishers,
P.O. Box 17, 3300 AA Dordrecht, The Netherlands.

Sold and distributed in North, Central and South America
by Kluwer Academic Publishers,
101 Philip Drive, Norwell, MA 02061, U.S.A.

In all other countries, sold and distributed
by Kluwer Academic Publishers,
P.O. Box 322, 3300 AH Dordrecht, The Netherlands.

Printed on acid-free paper

Printed in the Netherlands.

TABLE OF CONTENTS

Dimitri Ginev (ed.), Bulgarian Studies in the Philosophy of Science

PREFACE

In the last two decades the most important authors of Bulgarian studies in the philosophy of science - Professor Azarya Polikarov, Dr Sava Petrov, Dr Georgy Bratoev, Dr Stoyan Nikolov, Dr Atanas Danailov, Dr Georgy Gargov, and Dr Nedyalko Merjanov - passed away prematurely. They were many-faceted human beings, whose originality expressed itself in many different ways. We dedicate this volume to the memory of our beloved and distinguished friends and colleagues.

During the years in which this volume has been conceived and written, I have been sustained by the support of all contributors, most of whom are "working scientists". Our aim was to represent as many orientations of Bulgarian studies as possible. The initial project has been significantly improved as a result of our discussions. In addition to our contributions, I also include texts of Prof. Azarya Polikarov and Dr Sava Petrov, written in the 1980s. My belief is that both texts are representative of their authors.

I wish to express my deep gratitude to former Editor of "Boston Studies in the Philosophy of Science", Professor Robert Cohen, who encouraged me to initiate this project. I have a very substantial debt to Mrs. Jolanda Voogd (KLUWER ACADEMIC PUBLISHERS) for her dedicated cooperation. For the final preparation of the manuscript, special thanks go to two charming persons: Ms. Antoinette Koleva and Dr. Lilia Gurova. I am also much indebted to my friend and colleague Jassen Andreev. Conversations with him were instrumental in conceiving the general form of this project.

Dimitri Ginev

Dimitri Ginev (ed.), Bulgarian Studies in the Philosophy of Science

INTRODUCTION

There is no "Bulgarian school" in the philosophy of science. Yet there are "Bulgarian ways" of critical reception of ideas and approaches developed by Western schools. As early as the late 1930s, logical positivism gained ground in this country. Carnap's conception of the "formal mode of speech" and his claim that the philosophical investigation of science must be a logical analysis of scientific language were welcomed by the members of the Informal Group of Epistemological Studies in Sofia. Strangely enough, however, Bulgarian positivists were not inclined to accept the program of the unity of science. Because of the influence of the Diltheyan school in the first place, they remained champions of a fundamental distinction between the natural and human sciences.[1] They displayed a strong sympathy for Carnap's search for a "protocol language" providing an absolute foundation of the whole cognitive structure of the natural sciences. What they denied was the positivist attempt to represent the human sciences as systems of statements controlled by experimental verification. On their view, and this was actually a dominant methodological view at the departments of philology and history of the University of Sofia, experience in the human sciences is engendered by the interplay of researchers' theoretical concepts and the concepts embodied in the objects of inquiry. The objects of the human sciences are always (pre)conceptualized by their native interpreters and in their original historical contexts. This view was a specification of Dilthey's conception that in their pre-thematic status the objects of inquiry of the human sciences are already constituted within the triple relationship of understanding, experience (*Erlebnis*), and expression (*Ausdruck*). Bulgarian positivists from the pre-war period also subscribed to the Diltheyan conception that one has to relate every constituent of the cognitive structure of the human sciences to the historical-cultural dynamics in which "the whole of human nature" manifests itself.

After the communist invasion in Bulgaria (1944-46) the new political authorities proclaimed positivist (anti)philosophizing to be the propaganda of bourgeois ideology incompatible with the values of the Proletarian Revolution. The Informal Group of Epistemological Studies in Sofia was prohibited and its members were deprived of positions at academic institutions. The Diltheyan philosophizing had a similar destiny. Epistemology and methodology of science were replaced by a dogmatic discipline called Marxist-Leninist dialectics of nature and scientific knowledge, a discipline belonging to the canonized corpus of dialectical materialism. Ironically, this discipline created a niche in the 1950s for those who were looking for an escape from the ideological discourses of the repressive communist society. The very escape provoked an ongoing transformation (in terms of de-dogmatization) of the discipline. More and more the "dialectical-materialist studies of science" took on the form of Western philosophy of science.

Dimitri Ginev (ed.), Bulgarian Studies in the Philosophy of Science, ix-xii.

Professor Azarya Polikarov's (1921-2000) books - *Matter and Knowledge* (1961) and *Relativity and Quanta* (1963) - marked the culmination in this transformation. The former book opened the reception of post-positivist philosophy of science, whereas the latter initiated studies into philosophy of physics.[2] The interpretations of quantum mechanics played a central place in these studies. No doubt, Dr Sava Petrov's (1934-1989) investigations into the origin of the statistical interpretation of the wave function became paradigmatic for a whole generation of philosophers of physics in Bulgaria.

In the 1970s, post-positivism became fashionable among Bulgarian philosophers of science. A source of inspiration for them was first and foremost Norwood Russell Hanson's thesis that formal reconstructions of finished cognitive systems (e.g., planetary mechanics or classical thermodynamics) have to be replaced with the search for the rise of new patterns of explanation within the ongoing historical dynamics of the research disciplines. Due to Dr Georgy Bratoev's excellent work, the post-empiricist claim of the theory-ladenness of observations became entangled in the ideas of Wittgenstein's philosophy. It was this connection that provoked inquiries into the logical semantics of scientific languages.

The reception of post-positivist philosophy of science also owes an essential impetus from Professor Ivanka Apostolova's work on the notion of scientific style. The elaborations on this notion aimed at bridging Kuhn's historical relativism and Lakatos' quasi-Hegelian rationalism. To a certain extent, the spread of post-positivism was also a *sui generis* vindication of the pre-war "strange mixture" between logical positivism and the Diltheyan epistemology of human studies. More specifically, for many Bulgarian authors working in the 1970s, the post-positivist insistence on self-reflection on the history and explanatory practice of the sciences was reminiscent of Diltheyan historism. Yet the search for an essential distinction between the natural and human sciences was no longer on the agenda. The methodologies of cognitive historism and the models of scientific change promised new formulas of non-reductionist unification of the sciences.

To be sure, the orientation towards holist models of science's cognitive structure and dynamics was prompted by antireductionist research programs that dominated scientific life in Bulgaria. I mean programs based upon explanations in terms of "emergent properties and processes". Prominent examples in this respect are Methody Popov's school in cytology and general biology (whose research work is a typical case of what G. G. Simpson calls "compositionism") and Stransky-Kaishev's school in physical chemistry whose impact on domains like crystallography and colloid chemistry is still significant. (Subjects in the philosophy of science like "types of scientific explanation", "logical structure of scientific theory", and "the symmetry between explanation and prediction" are intcluded in natural science's curricula in the main Bulgarian universities since the mid-1930s.)

At the same time, however, the view began to emerge that post-positivist doctrines are in need of a Kantian framework. Many philosophers of science in Bulgaria displayed a disconent with the incommensurability thesis and the lack of a clear epistemological definition of scientific rationality. They believed that in supplying the post-empiricist approach to science's cognitive dynamics with a sort of transcendental epistemology, one would be able to find a remedy for overcoming cognitive relativism. From the works of the members of the Bulgarian Kantian Society, a new awareness of the pervasive role of the synthetic apriori emerged, affecting a whole cluster of traditional issues concerning the structure of scientific theory. Yet neo-Kantianism was not the only "Continental ingredient" of the studies into history and philosophy of science in Bulgaria. The shift in focus from the normative justification and rational reconstruction of scientific knowledge to reflecting upon scientific research in the making invited attention to linguistic and social-practical fore-structures of the research process. By implication, the ideas of French authors like Bachelard (especially his idea of a psychoanalysis of scientific intuition and imgination), Canguilhem (the social-historical origin of scientific norms), and Foucault dominated the scene in the early 1980s. Due to the growing interest in phenomenology and hermeneutics in the mid 1980s, ideas of Husserl's, Merleau-Ponty's, and Ricoeur's programs were "integrated" into the efforts of scrutinizing scientific thought. Gadamer's and Heidegger's variants of philosophical hermeneutics also came into play. An important additional reason for the willingness to accept Continental approaches to the philosophy of science is that analytical philosophy has never been a dominant tradition in Bulgaria.

A great diversity of tendencies flourished in the 1980's. Postmodern voices appealing for the replacement of the philosophy of science with social studies and cultural studies of science are still audible. At the opposite pole were attempts at a rehabilitation of logical positivism. Another tendency has been the growing interest in strategies for naturalizing philosophy of science on the basis of models developed in the cognitive sciences. Several programs of AI were employed in this respect. Analog devices without clearly defined rules were applied to the methodology of problem-solving in scientific research.[3] In trying to cope with the complexity of scientific research, other authors made use of Pylyshyn's account of a noninferential mechanism that mediates between the environment and the effect. In this context, the name of Professor Polikarov has to be mentioned again. Polikarov's position of "heuristic realism" was the best response to the situation of a growing diversity of conflicting tendencies. Leaning on his divergent-convergent heuristic methodology,[4] he succeeded in managing a "dynamic unity" in a variety of philosophical paradigms of science studies. Heuristic realism is a view predicated on (i) realism without Cartesian dualism; (ii) anti-foundationalism without "epistemological behaviorism"; and (iii) historical approach without historical relativism.[5] Although many aspects of heuristic realism are in need of further elaboration, it is successful in showing the way to surmount long-lasting dilemmas in epistemology that prevent one from reaching a genuine interaction between philosophical ideas and scientific theories.

In recent years, significant philosophical aspects have been incorporated in the works of Bulgarian mathematicians, logicians, and linguists. A leading exponent of the dialogue between philosophy of science and the formal sciences was Dr Georgy Gargov whose ideas (in modal logic, intuitionistic fuzzy logic, and mathematical linguistics) still have an immense impact on several lines of research.

To develop a map of all trends, positions, and thematic orientations of "Bulgarian studies in the philosophy of science" is, of course, beyond the scope of this introduction. Collected in the present volume are essays illustrating some of the aforementioned "Bulgarian ways".

Dimitri Ginev

NOTES

[1] About the "Diltheyan philosophizing in Bulgaria", see Ginev *Zwischen*.

[2] Three volumes of studies into philosophy of physics marked symptomatic stages in a *sui generis* development. The first one (Apostolova et al.) is devoted chiefly to the logical reconstruction of non-classical theories in physics. The problems of causality and determinism, the types of quantum logic, the justification of the law of conservation of energy, and special cases of interrelation between theory and experiment are the dominant subjects in this volume. The second one (Polikarov *Methodological Problems*) shifts the focus from the logical structure of theorizing in physics to the historical dynamics of the transition from classical to non-classical physics. The last volume (Tögel) puts more emphasis upon the cultural contexts and historical situations of emerging central ideas in contemporary physics. A similar evolution was also undergone by studies into the philosophy of biology, whose main exponents in the 1970s and 80s were Dr Stoyan Nikolov and Dr Atanas Danailov.

[3] See, for example, the contributions of Bulgarian authors in Bibel and Petkov.

[4] See in particular Polikarov, *Method.*

[5] For Polikarov's heuristic realism, see Ginev *Introduction*.

REFERENCES

Apostolova, Ivanka, Azarya Polikarov, and Boris Kandilarov (eds.). *Contemporary Physics: Tendencies, Methodological Problems.* Sofia: N.&I., 1973.

Bibel, Walter, and Boris Petkoff (eds.). *Artificial Intelligence: Methodology, Systems, Applications.* Amsterdam: Elsevier Science Publishers, 1985.

Ginev, Dimitri. "Introduction" in Ginev, Dimitri and Robert S. Cohen (eds.). *Issues and Images in the Philosophy of Science. (Scientific and Philosophical Essays in Honour of Azarya Polikarov.) Boston Studies in the Philosophy of Science*, Vol. 192. Dordrecht/Boston/London: Kluwer Academic Publishers, 1997: XIII-XVII.

Ginev, Dimitri. "Zwischen hermeneutischer Ontologie der Kultur und hermeneutischer Logik des Lebens." *Dilthey-Jahrbuch für Philosophie und Geschichte der Geisteswissenschaften* Band 10/1996: 246-253.

Polikarov, Azarya. "The Divergent-Convergent Method" in Robert S. Cohen, and Marx Wartofsky (eds.). *Methodological and Historical Essays in the Natural and Social Sciences. Boston Studies in the Philosophy of Science*, Vol. 14. Dordrecht: D. Reidel, 1974: 213-233.

Polikarov, Azarya (ed.). *Methodological Problems of Quantum Physics.* Sofia: N.&I., 1979.

Tögel, Christfried (Hrsg.). *Struktur und Dynamik wissenschaftlicher Theorien. Beiträge zur Wissenschaftsgeschichte und Wissenschaftstheorie aus der bulgarischen Forschung*, Frankfurt am Main, Bern, New York: Peter Lang, 1986.

Part I

INVESTIGATIONS IN THE GENERAL PHILOSOPHY OF SCIENCE

SAVA PETROV†

THE DANGER OF CATCHING NATURE IN CONTRADICTION

I say "danger", taking into account the common attitude towards contradictions and antinomies, but at the same time, as an old-fashioned dialectician and even more in tune with paraconsistent logicians, I make the implicit assumption that it would give a chance to the cause of Heraclitus – Hegel. This far-reaching philosophical hypothesis assumes that some of the contradictory and antinomic propositions we come across in rather immature scientific theories, are likely to be relatively objective truths. The concerned claims are very few indeed and mostly connected with self-reference or an intuitive expression of a state of change, but their interpretations in the sense of Heraclitus – Hegel would restrict the law of non-contradiction and would supply grounds for a "dialectics of the impossible".

It is relatively easy to come to the conclusion that the centuries-old dream of contradictory truths has become a fact on an abstract level, say in a set theory based on paraconsistent logic (Arruda, Priest) but as long as no empirical science is accustomed to paradoxes, Hegel's thesis still oscillates between Meinong's jungle and the world of sheer hypotheses. For the scientists, with few exceptions, there is not the slightest suspicion in the older than the classical logic axiom: what is real cannot contain contradictions (Plato). Strangely enough, the widespread "syndrome of Pygmalion", when one takes models for reality, cannot infect contradictions and paradoxes. The common attitude has made no progress since the ancient rejection of Protagoras' projection of contradictory properties onto reality. The easier it is to identify mathematical existence with consistency, the more difficult it is to understand a contradictory truth. What could that mean at all?

Both intuitive and logical notions of negation have left no place for "contradictory truth" except, maybe, in the case of sheer antinomy where the logic seems driven into a corner and intuitively things become even worse. Epistemologically, a contradiction cannot be thought of as a truth; ontologically it cannot exist as reality. Nevertheless, the boldest dreamers among dialecticians assume that every change and even mechanical motion do contain objective contradictions more or less similar to counterparts of logical contradictions and hence the perfect knowledge of motion and change should use contradictory truths, whatever that means. For some minds the understanding of becoming includes the truthfulness of "to be and not to be", though it seems irrational for the time being. A further step tries to restore the balance with the assumption that such a temporary

Dimitri Ginev (ed.), Bulgarian Studies in the Philosophy of Science, 3-15.

irrationality will evaporate in due time like the antipodes as an argument against the spherical form of the Earth.

A criticism of the alleged idea of Routley-Petrov that "micro-objects are inconsistent" (Smith, 110) provides grounds for reflection on this line of thought. The reference to a possible inconsistency of (or between) future theories that will represent the now unobservable and theoretically unknowable by quantum mechanics objective motions of micro-objects in the paper of R. Routley, (1979), is much more cautious than the startling phrase "micro-objects are inconsistent".

In principle, I find it impossible for a philosopher, staying on his own level, to assert or to negate any empirical claim in a conclusive way. By the way, Locke grasped this impossibility somewhere in 1687!

My point has been, and still is, that some peculiarities of quantum-mechanical description of the micro-world could be used as modest suggestive considerations for the plausibility of the old hypothesis that in exceptional situations, say, in a state of motion, along with consistent truths, there could also be valid contradictory ones. Obviously, such a heuristic suggestion could not be verified by contemporary macro-experiments and macro-phenomenological theories. J. Smith's conclusion that the Routley – Petrov claim is "empirically false" shows a misunderstanding. Such hypotheses are unverifiable at present. Only micro-micro-experiments, unthinkable for the time being, could go as far as to prove or disprove a hypothesis about the objective motions of micro-objects.

As a matter of fact, we have only one straightforward, and hence improbable idea of a possible objective counterpart of a logical contradiction. It comes to a direct objectivization of the logical characteristics of a contradictory claim in the form of "to be and not to be" as Graham Priest has put it. However, the ontological referents of logical laws are not bound to be as simple and straightforward as that. The diversity of the possible objective counterparts of logical inconsistency of formal systems and empirical theories should be rather relatively comparable with the diversity of ontological referents for the logical consistency of our claims and theories. Now we have to use phrases like "it is and it isn't" more or less as definitions of a state of change or, if you want, as a sign of intuitive protest against the idealizations in use.

We should be very careful not to turn such a priori general phrases into specific proposals for premature conclusions of current scientific problems!

Another warning concerns the expected wider use in micro-physics of purely theoretical terms and claims that have no objective referents and that are practically unavoidable in a theory possessing objective validity. Imperceptibly we liberalize the use of the notion of truth for claims built out of pure theoretical concepts. These claims, however, are not truths in the strict sense of the correspondence theory of truth. Speaking of contradictory truths one suddenly forgets about this liberalization and starts looking everywhere for strictly objective referents. We should discard such restrictions by definition. We have to be as liberal to contradictory truths as we are to consistent truths, when asking ourselves about objective referents. Maybe the complementarity principle is only the first swallow for a future series of mutually inconsistent descriptions of micro- and mega-worlds, too complicated for our crude brains. If our correspondence theory of truth were dialectical enough, we should have no ideological problems either with consistent descriptions of inconsistent reality, or with inconsistent descriptions of consistent reality. However, one ought to find specific explanations of these unusual relations in every concrete instance. The inconsistent descriptions of inconsistent objects appear to be a forbidden land of unlimited impossibilities. For rare romantics, however, it is attractive because of its mystery.

According to Folse, N. Bohr himself, understanding matrix mechanics in classical manner, supposed in 1925-1926 that "nature commits a contradiction" (Folse, 63-64), or rather expected that nature would be caught in a contradiction, so impressive were the experiments showing the wave properties of electrons. But quantum mechanics with its probabilistic interpretation (1927) found a fantastically simple way out of the paradoxes of wave-particle duality and physicists had been calmed because they were reduced to complementarity and not to contradictions. However, some bitterness remained: why can't we understand the deeper laws of objective micro-reality without falling into contradictions concerning trajectories (Dirac); why is there no final escape from the logical opposites of "either – or" (classical mechanics) and "and – and" (wave mechanics) and must we constantly balance between them (Schrödinger)? These problems are still with us. In 1786 the Academy of Science in Berlin announced: the Academy desires an explanation of how it is possible that so many correct theorems have been deduced from contradictory suppositions. In a sense, Cauchy and Weierstrass have solved this problem by finding consistent presuppositions of mathematical analysis. Nowadays the wide-spread implicit question sounds like: why is it forbidden by so many theorems under the threat of contradiction to say anything about the objective motion of micro-particles?

Philosophers are not easily at peace with their realistic conscience according to the "tranquilizing philosophy" (Einstein) of the Copenhagen interpretation. Even the exceptionally convincing validation of quantum mechanics by a routine experiment did not dispel any doubts concerning the strange rationality of the really real unobserved motion of micro-objects, which evades any consistent description. Apparently quantum mechanics is not non-classical enough for it.

We explain everything by interference of amplitudes of probabilities of future experimental results but the now supposed objective or half-objective counterparts of these strange amplitudes are potential possibilities that "cannot kick or be kicked". Well then, what is it and how does it interfere with itself and with slits in experimental devices stretched out a hundred times the electron's size?

Loui de Broglie's "double solution" with real particle and/or real wave and many other concepts have been on the scientific stage for 60 years with big promises and no success. After two generations of unsuccessful efforts to reconcile quantum mechanics with scientific realism, maybe the time has come for more exotic attempts, and the ancient dialectical hypothesis that the state of change should contain objective contradictions is ready for a risky methodological prediction like this: a description of the now unobserved objective motions of micro-objects "could be obtained in the future either by simultaneous utilization of two theories, alternative to each other, or by means of a theory based on non-Aristotelian logic, i.e., logic where contradictions could be true (Petrov, *Logical Paradoxes*, 386), (Petrov, *Hegel's thesis*, 75-76). I have not touched upon these problems for many years, excepting for when N. da Costa, G. Priest or F. Asenjo made their short visits to Sofia, but I will try to gather some heuristic considerations that apparently increase the verisimilitude of this vague orientation.

First, it should be stressed that it is not necessary just now to have a passable logical and/or scientific substantiation of the old idea to express the state of change with contradictory sentences in order to preserve it as a philosophical hypothesis. Leibniz's criticism of Newton's conception of space was an a priori hypothesis for some 300 years but in the end the relativity theory took the place of Leibniz. Democritus atomism found its scientific realization at the beginning of the 19th century. We can find good company for the thesis of Heraclitus-Hegel. We need to remember such possible precedents because philosophers do have uneasy feelings about this and other cases of contradictions between philosophical and scientific conceptions, as if it were a life-and-death question.

Even worse, accepting the rule "science is always right", one is ready to invert the ontological relations between the actual and the potential, and to upset everything in order to do away with any shadow of difference with science. The consequences of such misunderstood "scientification" of philosophy promise to be as bad as the former "philosophication" of science. I wonder whether it is not already time for an inversion of Newton's "Physics beware of metaphysics" to "Philosophy beware of science".

Scientific philosophy (including dialectical materialism) is expected to produce far-reaching hypotheses and it is quite natural for some of these hypotheses to have no logical or scientific grounds and/or to be even in contradiction with today's data.

"Philosophy and science are in the same boat" (Quine). Yes, but philosophers watch the horizon and it is excusable for them to cry "Land" sometimes at the sight of a mirage. While scientists fish from the stern and are used to turning their back to high principles as far as they do not need them for practical purposes. The relations between Heraclitus and Aristotle, Hegel and Russell, etc., coming now to G. Priest and H. Wessel, appear to be of the same type.

Practice, science and logic say in one voice: a change of something means to be at different times in different states, and that's all; the passage as an inner property is a myth (see Williams, Ajdukiewicz, Sinov'ev and Wessel). However, some philosophers are dissatisfied with this plain truth, and trusting their intuition more, they stubbornly repeat: such descriptions, never mind how useful they are, reduce change to a sum of states in rest or to a sum of elementary changes. Our intuition needs – in addition to these consistent descriptions – a state of change, even if it should be expressed by contradictory statements. And here comes Hegel: something moves not because at one moment of time it is here and at another – there, but because at one and the same moment it is here and not here; the dialectical synthesis slips out from "it is... and it is not" (Hegel, *Werke,* Bd. I), originating new notions in a consistent new theory in which the contradiction has been sublated, that is, overcome and preserved (Bogomolov). This, however, is a lonely position met with no sympathy even among dialecticians (Havas). The few bold logicians who have tried explications of the state of change by contradictory claims find great difficulties (Priest). Nevertheless, non-classical theories like quantum mechanics and paraconsistent logic, directly or indirectly, have made some hesitant steps towards this dangerous zone, where objective dialectical contradictions partly coincide with ontological counterparts of logical contradictions.

It should be stressed that the hypothesis of Heraclitus-Hegel is bound neither to dialectical ontology nor to paraconsistent logic, though it could be added to both with essentially new consequences. By the way, dialectical logic in its initial historical sense has not been conceived as a logic in the proper sense of the word at all. Dialectical logic in dialectical materialism and other, similar philosophies is a methodology, a normative application of a dialectical world-view when the norms have only heuristic value. Dialectical logic (= methodology), like any other theory, can be formalized, at least partly, on the ground of classical logic, if no assumption like Heraclitus-Hegel's hypothesis is added or on the ground of non-Aristotelian (paraconsistent) logic, if it is added.

Some paraconsistent logicians (da Costa and Wolf) do have dialectical motives and do try to formalize fragments of dialectics through paraconsistent logic; others (Goodman) have taken into consideration inner logical motives as the possibility of a logic that is dual to the intuitionistic getting on with contradictions. In any case, the intuitive semantics of the contemporary paraconsistent systems, especially their negations, is not very attractive and the most cautious position would be to consider the present non-Aristotelian logic as an abstract guarantee rather than a conclusive realization of the possibility for logical validity of contradictory truths.

Secondly, I would reproduce the general heuristic approach, applied by me to every logical, semantical, mathematical, physical or any other paradox, in order, if not to check, at least to increase or to reduce the chances for further and more specialized considerations of it as an appearance of an objective dialectical contradiction.

We do not know much, if anything, about the possible objective meaning of the paradoxes in our knowledge, but the methods of their solutions, though also under discussion, are well known in many respects. That is why we have to look for some indirect links between the solutions and the supposed objective contents of the paradoxical claims. As a matter of fact, we are interested in one thing – are there any objective contents at all or not, in other words, is a given paradox a relative truth or an error in the full epistemological sense and not only in the narrow professional one.

To put it mildly, the criteria I have used are not one hundred percent reliable but as far as I know, nobody has suggested a better one.

In a well-developed theory the truths build up a unity. One cannot eliminate a single truth from the whole theory without essentially encroaching on some other truths. Just now we are having a bitter sociological experience with this very property of truths.

The elimination of an error, or even of a deep fallacy, should be a quite different thing, although serious mistakes have their own roots in knowledge or in reality.

The notions, assertions and theories that we put in the place of eliminated errors, if adequately formulated, should be connected with the former roots of the errors much better than the notions they replace; and, if possible, under given circumstances, such paradox-error resolving notions should have objective meaning of their own, being empirical or theoretical terms but not purely theoretical ones.

And, on the contrary, a paradox with a hidden truth-value should be eliminated through methods, notions, assertions and theories, which (1) either have a non-logical character, i.e., with the given paradox they also reject other, irreproachable parts of the theory, or (2) are themselves pure constructs without any objective contents, or (3) somehow encroach upon the completeness of the respective theory.

Naturally, such unpleasant features could be also explained by other causes and not necessarily by a coincidence of a logical contradiction and an objective dialectical one. For that reason it is with the indications of plausibility that we are concerned here, not the criteria for the truthfulness of the Heraclitus-Hegel's cause.

In order to grasp the third consideration, let us now come back to the paradoxes of the wave-particle duality of micro-objects in their intuitive realistic understanding, paying attention to the peculiarities of the methods of their quantum-mechanical solutions.

Under the conditions of an idealized experiment, there are a priori two possibilities for the motions of unobservable micro-objects as real substantive entities passing through the slits A and B:

(a) Every micro-object passes either through A and only through A, or through B and only through B;

(b) Every micro-particle passes simultaneously through A and B.

If classical conceptions of wave and particles or arbitrary non-classical but realistic (objective) ones are assumed, the Aristotelian logic takes upon itself to guarantee the alternative character of these two possibilities. The simultaneous realization of (a) and (b) would be equivalent to an objective paradoxical situation and this is what N. Bohr was afraid of in 1925-1926. Light-waves have shown properties of particles and matter-particles have shown properties of waves. How could it be possible for one and the same object to spread everywhere in space as a wave and to interact at a point as a particle without detriment to the consistency of the world?

However, Nature shows itself smart enough to escape the paradoxes on the macro-level at least, for (a) and (b) are realized only in different experiments and with different devices, although with the same micro-objects. If we keep strictly to the empirical data and their mathematical connections, without any thought of the objective reality prior to any macro-measurement, the paradoxical aspects of the simultaneous realization of (a) and (b) could not reach the logical exclusion of contradictions.

This has been enough to change radically the face of physics from classical to non-classical. Considering waves (amplitudes) of probabilities of receiving experimental data instead of real entities, quantum mechanics explicitly rejects any claim on the objective situation prior to measurement. One, however, implicitly accepts that an unobserved micro-object potentially goes through both slits but actually does not pass through any slit, giving, nevertheless, as much interference as a real wave.

If only we try to give an objective description of unobserved motions that are half quantum-mechanical, logical contradictions would be with us once more. The contradiction enters when one demands a complete objective description of the phenomena (Kramers). The dynamic properties of micro-objects, unobservable by macro-devices, are not "Golden mountains". In the end, we must have an explanation of this peculiar restriction of our knowledge or of nature's capacities. Unthinkable as it is, the assumption of contradictory truths has to stay among the "possible" methodological grounds of such explanations until the moment of enlightenment. The quantum-mechanical solution of the paradoxes of wave-particle duality is achieved at the price of strict renunciation to describe the actual objective behaviour of the non-observed micro-objects: no elementary quantum phenomenon is a phenomenon until it is a registered phenomenon (Wheeler).

The main device for this purpose is the probabilistic interpretation of the universal mathematical tool – the Ψ-function, which is another word for the amplitude of probabilities. Instead of answering where a micro-object actually passes through, quantum mechanics says that there is a probability P1 for passing through slit A, and probability P2 for its passing through B. Thus there remains an unsolved problem for today's methodologists and for the physicists of tomorrow: how does micro-nature achieve what we achieve by interference of amplitudes of probabilities.

Why does Nature demand that we walk along a logical tightrope if we wish to describe her (Feynman-Hibbs)? It is more than improbable that the actual existence of fundamental dynamic qualities of elementary particles depends on such a subsidiary detail as man's observation. Then what?

Scientists still repeat after Aristotle: a Heraclitean (a Hegelean, a dialectician, etc.) may deny the principle of non-contradiction in words, but he is not actually thinking that way (Aristotle, *Metaphysics*). Since 1927, however, a scientific theory has cast a shadow of doubt that itself escapes objective contradictions in words and the real motions of micro-particles proceed irrationally. An indivisible particle cannot show wave-like behaviour without being at once here and not here. The size of a free electron is about 10^{-13} cm. The same electron shows wave-like properties to 10^{-3} cm! What should the inner nature of motions of micro-objects be like in order to appear on the macro-level in complementary pictures?

Neither the potential possibilities (London and Bauer), nor Heisenberg's tendencies or Popper's propensities give us any hint about the reality, which should explain the success of the amplitude of probabilities as a measure of any kind of changing micro-properties. We should keep our mind open to anything: Nature could always surprise us.

Fourth, if any attempt to describe what should happen in the objective micro-reality, under the conditions of the two-slit experience when no observation devices are used, leads to contradictions (Heisenberg), it might appear tempting to jump to the conclusion that the dynamics of micro-objects is objectively antinomic and that is why quantum-mechanical descriptions must be probabilistic and hence incomplete, and this is how it saves itself from contradictions. Obviously, such an explanation would be logically erroneous, for the contradictions in our thoughts usually have causes that have nothing to do with any objective counterparts of logical contradictions.

The only right we have is to make a heuristic suggestion that maybe the objective micro-reality contains the counterparts of logical contradictions and that is why, bound to the law of non-contradiction, we refuse any description of the objective state of motion of micro-objects prior to micro-experiments. However, the innate belief in the consistency of Nature is so strong that no one happens to use this theoretical right in practice.

Just here, to fill up a heuristic gap, my claim finds its place: we should use, among many others, the hypothesis that prior to any macro-experiences there must be some actual and not only potential motions of micro-particles even, at the worst, antinomic ones. Nothing more than that – an abstract possibility for speculations until one waits for future micro-micro-experiments to decide the question. A static objective counterpart of logical contradiction allowed by a non-Aristotelian logic does not say much ontologically and the next abstract suggestion might be connected with positive and consistent characteristics of the motion of micro-objects unobserved by macro-devices, except the initial "to be and not to be", say, wave-like and particle-like motion.

The additional considerations that one could gather to increase the verisimilitude of the fantastic conjecture of antinomicity of some objective characteristics of the micro-world, could not change essentially the logical status of the situation. But they are curious for a reiteration here, maybe.

1.

Quantum mechanics is the third example where a withdrawal from the actual into the much wider field of the potential is used to save a theory from paradoxes. The other two cases are the Cauchy – Weierstrass solution of infinitesimal paradoxes and Boltzmann's statistical interpretation of the second law of thermodynamics, which attenuated the contradictions connected with irreversibility.

Moore (I mean the mathematician E. X. Moore, not the philosopher G. B. Moore) exaggerates in supposing that the existence of an analogy between different theories gives grounds for a new theory unifying the initial theories, but the analogy under consideration gives at least a hint that the solutions of some paradoxes are somehow related to a substitution of the actual for the potential.

For mathematicians this cuts no ice but physical theories appreciate very highly the description of actual reality and they must be constrained by extraordinary circumstances to change it for a description of potential possibilities.

The paradoxes themselves are the most obvious common feature of the above mentioned three remarkable examples from the history of science. In this way our conjecture that actual motions are somehow intrinsically connected with objective counterparts of logical contradictions scores.

2.

The next point comes from the specific logical force used by the formalism of quantum mechanics to reject in principle the possibility of any future introduction of hidden variables in quantum mechanics itself or in any other theory of the micro-world, which would respect its truthfulness. That beats everything. No other theory dares to prove logically that its description of reality is complete! An elementary feeling of self-reservation should suggest to the realists that they must not oppose the existence of the objective world to the well-substantiated completeness of quantum mechanics.

What realists really mean is that human knowledge of the motion of micro-objects temporarily reduced to quantum mechanics is surely incomplete, simply because we have only macro-projections of it and theories that are essentially macro-phenomenological. Quantum mechanics itself could be complete and may stay like that, but human knowledge about the micro-world will develop.

Usually, realists do not go into such details and being brave men they fall one after another in the 60-year battle for hidden variables, invented here and now, without any experimental suggestions against impeccable theory and practice. One of the many culpabilities for the useless victims is the belief in the principle of correspondence.

The mathematical (von Newman, Gleason, Kochen – Specker, Bell, etc.) proofs of the impossibility of hidden variables as well as the experimental ones are based directly or indirectly on the supposition that a theory with hidden variables, be it quite new or a transformation of the present one, should be able to repeat the quantum mechanical results confirmed experimentally, that is to say, one takes for granted the validity of the principle of correspondence in its generalized form.

Hence, the hope for hidden variables involves the acceptance of their non-locality, which is a desperate hope for realism. A non-local theory with hidden variables would hardly be better than quantum mechanics. The action at a distance is as disgusting to materialists as the interference of probabilities. However, nowadays (after Aspect et al), the non-locality in somewhat stronger form as non-separability of micro-events is an experimental fact. And somehow noone remembers the classical approach of the philosophers: "so much the worse for the facts". Nevertheless, as a kind of "Zukunftsmusik", such an approach has its chances. If we suppose that the quantum-mechanical knowledge is as restricted as the sensory knowledge of secondary qualities, the expectation of future possibilities for micro-micro-experiments would appear to be the only possibility for a break-through the wall of contemporary macro-projections. In an axiomatic theory of sensory knowledge one would surely prove theorems of the impossibility of hidden variables, i.e., of primary causes of colors. The discovery of experiences with devices for which dispositional properties like colors do not emerge, has put things in their place. However, the deduction of color vision from the theory of the primary causes of colors remains an impossible one. As a rule, neither the one-sided knowledge of dispositional qualities, nor the one-sided knowledge of substantative ones permits a simple deduction from the former to the latter and vice versa. Very likely, a similar situation will repeat itself in the relations between the present quantum mechanics based on macro-morphous data and the eventual theory of actual objective motions of micro-objects based on micro-micro-experiments, for which macro-dispositions do not emerge.

In my opinion, we could see a twinkling light in the darkness using the analogy with secondary qualities and discarding the correspondence principle in the relations between quantum mechanics and a fundamental future theory of objective motions of micro-particles.

This would mean that we should expect a pair of mutually exclusive alternative theories of the dynamics of the micro-world: a new substantive theory and the present quantum mechanics. We could then calmly see the theorems and the facts proving the impossibility of hidden variables on the level of macro-phenomenological experiences and anything built on them. Such theorems would be quite natural for theories that are mutually contradictory although separately consistent and without any hindrance to the necessity of hidden variables on philosophical grounds. In the worst case, the substantive theory itself could be based on non-Aristotelian logic, which allows contradictory truths.

3.

The labyrinth of quantum logic has been criticized "en bloc", so to say, "from the right" with arguments that are expected to show that the two-slit experiment has not proved the non-distributivity of the logical operations implicitly used in quantum mechanics (Selleri, etc.). Criticisms "from the left", say, in a sense, that one should go a step further and invalidate the law of the excluded middle (Gardon), are rare exceptions and are easily discarded from the standpoint of orthodox quantum theory. With respect to the principle of observability, unusual for a formal science, logicians, as far as I know, did not try to guess what could be the logical structure of the ontology of the unobserved motions of micro-objects, though in the 30s it was the first thought of the philosophers who suggested the three-valued logics as an escape from the paradoxes of wave-particle dualism. Quite independently of such an interest, which was theoretically expected and historically unrealized. Since the 60s the development of non-classical logic has included more and more non-Aristotelian (paraconsistent) systems. Some of them could be used in principle as a ground for fragments of possible theories of the actual and now unobservable motions of micro-objects. The present systems of paraconsistent logic have some particularities, especially with the properties of negation, which do not make them a very attractive means for the purpose. Hower, the bad luck with the attempts for alternatives of the quantum theory promises enough time for further development of suitable logical backgrounds.

It is quite natural that the fears or the hopes of catching nature in contradiction, say, in 2226, might be as useless as they were in 1926. The micro-world may appear illogical not because it is illogical in itself but because we ascribe to it macro-properties, which the micro-objects do not possess. If in the long run we consider the micro-world as it is in itself, then its logic would appear to be the common and its illogicality would disappear (Aronov). A single contradictory truth in an empirical science would mean such a great change in man's outlook that we must take into consideration this possibility, never mind how small it is.

4.

N. Rescher and R. Brandom in their monograph *The Logic of Inconsistency* (1979) have considered Everett's (1937) many-worlds reconstruction of quantum mechanics as "an invitation, though not an irrevocable demand of a logical apparatus that is inconsistency-tolerant" (Rescher and Brandom, 60). Apparently, no one has yet used this invitation or has even recognized it, excepting the pair Rescher – Brandom. It has also been suggested that the unknown causes of the Heisenberg uncertainty principle have something in common with ontological inconsistencies in motion (Priest, 225) but such suggestions have not yet found effective continuations in physics. So far, without any continuation, there remains V. Gerard's attempt to build an algebra of two-valued logic that transforms itself into a quasi-dialectical logic (A & $\bar{A}$ is never true) under certain conditions and gives the theory of probability; under different conditions it transforms itself into a dialectical logic (A & $\bar{A}$ could be true) and gives the quantum theory of probability. In the quantum theory of probability the so obvious classical assertion that there should be a joint probability distribution for an arbitrary pair of physical variables is not held.
If we, like to have a spectrum of the future developments of microphysics that is as rich as possible, we should not leave unattended similar bunches of exotic and marginal opinions.

My last comment, physicists say half-jokingly that in the quantum domain an idea should be crazy, irrational from the standpoint of everyday life, in order to be true. Well, for centuries the hypothesis of contradictory truths has had the fame of an irrational idea and on this basis it should fit into the philosophy of quantum physics. If the contest from Heraclitus to Hegel and to our day has not much impressed physicists, the general situation should be blamed.

Current ideas pass such an ironical criterion with the same facility. The interference of sheer possibilities cannot obviously have straightforward objective counterparts and no one has the slightest idea why it works at all. The micro-object – macro-device physical interaction was an acceptable explanation for Heisenberg but in the course of time it lost much of its epistemological force and now one uses its modifications more or less as formal excuses. The causeless emergence of the dynamic properties of micro-objects just at the moment of macro-measurements would only be worse. The conception of a primary mechanismless wholeness of the micro-world or at least of any two micro-objects that had been in interaction somewhere in the past, is coming now in vogue with David Bohm's easy hand and very likely it would appear to be a new version of the "Pygmalion syndrome", i.e., when one takes epistemological models for objective reality. Quantum-mechanical wholeness has become a fund for easy parallels with eastern mysticism (Restivo).

Under normal circumstances one would hardly put in use the most unbelievable hypothesis – an objective antinomicity; but driven into a corner by the tandem Bell – Aspect, realists would not hesitate much before remembering Hegel's challenge to classical logic side by side with the idea of a fundamentally inseparable world. After nearly a century-long effort to understand the nature of micro-objects in all logically possible ways, one still faces the following alternatives. The micro-objects are:

(1) corpuscular but not wave-like; (2) wave-like but not corpuscular; (3) neither wave-like, nor corpuscular; (4) wave-like and corpuscular simultaneously but in different relations. Maybe it is time to dare to think about the unthinkable, i.e., to assume that micro-objects are (5) wave-like and corpuscular simultaneously in one and the same relation. This philosophical liberty could hardly be put to use immediately but it appears to be the most non-classical background for the further development of physics that one could imagine.

REFERENCES

Ajdukiewicz, Kasimierz. "Fragen der Logik." *Deutsche Zeitschrift für Philosophie sonderheft* 3 (1956).

Aronov, R. "K probleme prostranstveno-vremennih otnoshenij v kvantovoj phisike". *Voprosi Filosofii* 4 (1984).

Arruda, Ayda. "A Survey of Paraconsistent Logic". In: Arruda, Ayda I., R. Chuaqui, and Newton C. A. da Costa, (Eds.) *Mathematical Logic in Latin Ametica*. Amsterdam: Elsevier North-Holland, 1980.

Aspect, Alain, J. Dalibard, and G. Roger. "Experimental Test of Bell's Inequalities." *Physical Review Letters* 49 (1982).

Bell, John. "On the Einstein – Podolsky – Rosen Paradox." In: Wheeler, John A. and Woiciech H. Zurek. (Eds.) *Quantum Theory and Measurement*. Princeton, N.J.: Princeton University Press, 1983.

Bogomolov, A. "Dialectical Contradiction and its Solution." In: *Philosophy in the USSR*. Moscow: 1977.

Bunge, Mario. "The Einstein-Bohr Debate over Quantum Mechanics: Who Was Right about What." *Lecture Notes in Physics* 100 (1979).

Costa, Newton C. A. da and R. Wolf. "Studies in Paraconsistent Logic I. The Dialectical Principle of the Unity of Opposites." *Philosophia* 9 (1980).

Dirac, Paul. *The Development of Quantum Theory*. New York: Gordon and Breach Science Publishers, 1971.

Everett, H. "Relative State Formulation of Quantum mechanics." In: Wheeler, John A. and Woiciech H. Zurek. (Eds.) *Quantum Theory and Measurement*. Princeton, N.J.: Princeton University Press, 1983.

Feynman, Richard and Albert Hibbs. *Quantum Mechanics and Path Integrals*. New York: McGraw Hill, 1965.

Folse, Henry J. *Philosophy of Niels Bohr*. New York: North Holland, 1985.

Garden, Rachel W. *Modern Logic and Quantum Mechanics*. Bristol: A. Hilger, 1984.

Havas, L. G. "Differences in the Unity." *Logique et analyse* 29 (1986).

Hegel, George W. F. *Werke, Bd1*. Berlin, 1832-87.

Heisenberg, Werner. *Die physikalischen Prinzipien der Quantentheorie*. Mannheim: Bibliographisches Institut, 1958.

Kramers, Hendrik A. *The Foundations of Quantum Theory*. Amsterdam: North Holland, 1957.

London, E. and E. Bauer. *La théorie de l'observation en mécanique quantique*. Paris, 1939.

Petrov, Sava. *Logical Paradoxes in Philosophical Interpretation*.Sofia: Nauka I Izkustvo, 1971. (In Bulgarian)

Petrov, Sava. "Hegel's Thesis of Contradictory Truths." *International Logic Review* 17-18 (1978): 69-76.

Priest, Graham. *In Contradiction*. Dordrecht: M. Nijhoff, 1987.

Rescher, Nicholas and Robert Brandom.*The Logic of Inconsistency*. Totowa, NJ: Rowman and Littlefield, 1979.
Restivo, Sal P. Parallels and Paradoxes in Modern Physics and Eastern Mysticism. *Social Studies of Science* 8 (1978).
Routley, Richard. "Dialectical Logic, Semantics and Metamathematics." *Erkenntnis* 14 (1979).
Selleri, Franco and G. Tarozzi. "Why Quantum Mechanics is Incompatible with Einstein's Locality?" *Physical Letters* A 119 (1986).
Sinovev, Alexander. *Logical Physics*. Dordrecht: M. Nijhoff, 1985.
Schrödinger, Erwin. "Die gegenwärtige Situation in der Quantenmechanik." *Naturwissenschaft* 23 (1935).
Smith, J. "The Consistency of the World." *Erkenntnis* 24 (1986).
Wessel, Horst. *Logik*. Berlin: Springer Verlag, 1986.
Wheeler, John A. and Woiciech H. Zurek. (Eds.) *Quantum Theory and Measurement*. Princeton, N.J.: Princeton University Press, 1983.
Williams, D. "The Myth of Passage." *Journal of Philosophy* 48 (1951): 457-472.

VIHREN BOUZOV

SCIENTIFIC RATIONALITY, DECISION AND CHOICE

1. SCIENTIFIC RATIONALITY

In my view, without a clear-cut understanding of the concept of rationality itself and of scientific rationality in particular, it would not seem warrantable to take in science as a pattern of rationality, and to consider scientific activity as being more rational than other types of human activity. Else, its consideration as an abstract construction with dogmatic and restrictive characteristics – or as a void of content and wholly evaluative concept – will, for sure have serious grounds. Indeed, it is of major importance to find out positive solutions to problems of the nature of rationality in the context of the present existing intellectual crisis, when the criticism of science and irrationalism are in an aggressive offensive and there is talk about 'collapse' of scientism and foundationalist programs in philosophy and of the scientific perception of the world as well (Tuomela, *Science*, 93).

Herein I will try to set out certain innate traits of scientific rationality, by means of making a comparison between leading subjective and objective accounts of it in aspects representative for their explanatory potential.

Scientific rationality might well be taken in as a system of specific norms, originating from, and upheld by, a scientific community; norms offering a choice of best decisions in a set of rival alternatives. Hence, a study may be developed up to the evolvment of a uniform conception of scientific rationality and its variants.

The concept of rationality relates to the instruments of carrying out human activity and suitability in terms of aims. Classical philosophical tradition draws a line of demarcation between rationality of thinking and rationality of action. The former spells out universal laws of the Reason guiding nature, society and knowledge (Toulmin, *Cosmopolis*). This type of rationality is uniform for all people; it does not depend on time and social conditions. It characterizes the development of thinking itself, not the development of reality. The rationality of thinking is an emanation of transcendental Reason; typically it is identified with the laws of logic and other 'innate' truths. The rationality of action is determined by aspects of: situation of choice, limited ability and knowledge of the subject, and free will. These aspects are rational, falling in with aims, and conductive to their scoring.

At present, philosophy stipulates an elimination of the difference between thought and action, theory and practice. Thought itself is a type of practical activity, a singling out of alternative decisions. The subject's development is a process of a

Dimitri Ginev (ed.), Bulgarian Studies in the Philosophy of Science, 17-29.

nature determined by internal and external factors. The distinction made between methodological and practical rationality, between inferential and behaviourist conceptions of scientific reasoning, stems from the unjustifiable thought/action opposition.

Positivist and postpositivist philosophy of science identifies rationality via a set of methodological rules. This conception of rationality presupposes construction of an universal method and systematization of sciences. Scientific theories have to keep up to certain rules and standards, the emanation of logical severity. Rationality is guaranteed by means of abidance by such rules and standards, themselves an expression of the procedures of acceptance, justification and criticism of knowledge. Their uniqueness and logical power determine the priority of science vis-à-vis other forms of knowledge. Those rules are means of gaining an objective, true knowledge or an adequate explanation of phenomena. Its explication leads to the construction of rational models with claims on revealing the nature of scientific change (Newton-Smith, *The Rationality*, 17).

So far, the philosophy of science has not been successful in proving convincingly that rationality of scientific knowledge might be perceived of as keeping up to rigid methodological rules. P. Feyerabend thinks that such a rationality is a holdback in the real advance in science; it imposes limitations on human freedom. Scientific progress makes headway through breaking up the constraint of methodological rules (Feyerabend, *Against Method*). The hope that such general and all-embracing directives exist has been dwindling away without let-up, due primarily to the impact of the established pluralism of the forms of rationality. Feyerabend convincingly points to the real variety of 'rational' standards. On their part, the latter determine different cognitive strategies and practices. One might rightfully infer that the interpretation of a certain cognitive procedure as rational could not be pared down to a finite set of characteristic feature qualities. The concept of rationality is of a relative and changeable nature. There exists no idea (or activity or tradition) that might be assessed as 'the one-and-only rational' alone, for good. R. Rorty works out this entirely justified conclusion of Feyerabend to an extreme relativism (Rorty, *Philosophy*, 331), an assertion that is completely unacceptable, since there are evaluative and normative invariants going to the making of rationality as well.

Are scientists rational in terms of their methodological conception of rationality? L. Bergström is right in saying that it 'confuses means and ends, or process and product, in a certain way'. Methodological rules could be perceived of as forwarding some of the aims of science, not as determining any particular behavior of individual scientists (Bergström, *Some Remarks*, 1-3). Science needs no fixed, tight rules of a method; rather, it needs objective criteria in the effective selection of aims, and means of their achievement. This fact should be a guide in the recovery of the until recently neglected problem of scientific discovery. The methodological conception of rationality ignores values with key roles in a practical choice.

The decision theory is the most successful claimant in the evolvement of a model of practical rationality. In terms of practice, rationality is a choice determined by good grounds (Shick, *Making*, 34). The theory of decision seeks to offer a plausible model of rational action and to formulate general principles of rationality guiding decision-makers under conditions involving risk and unreliability of information. The agent has to make a choice in the presence of several alternatives: their results depend on the actual occurrence of a situation – in a set of situations mutually excluding each other. The agent will be striving to act in a way that might bring about the maximum meeting of his wants or preferences. The choice is rational if it maximizes the expected utility of action (Jeffrey, *The Logic*).

The possibility to use the sophisticated mathematical constructions of the decision theory as a means of analyzing of scientific knowledge is recognized and an expanded debate going on (Howson, Urbach, *Scientific*; Maher, *Betting on*). It is an incontestable fact that scientific researchers act, as a rule, in a state of inadequate availability of tools and incompleteness of information. The choice of hypotheses in a set of competing alternatives, coached to theoretical norms of decision-making and based on empirical tests, is usually viewed as a model of inductive knowledge. C. Hempel assumes that the problem of application of inductive logic in the formation of rational beliefs can be treated as a special case of the more general 'problem of formulating rules for a rational choice or decision in the face of several alternatives'. Scientists act rationally when they seek to maximize certain 'purely scientific, or epistemic utilities' on the basis of beliefs accepted by them (Hempel, *Aspects*, 73, 76). It becomes clear, in the development of further discussions, that the theory of decision embraces, completely or in part, numerous questions of relationships between probability and induction. Yet, it would seem that – irrespective of its, to a certain extent, impressive applications to the topics of confirmation, statistical inferences or economic behavior – P. Suppes's pessimistic conclusion that it has not helped us to understand the concept of rationality and to describe a rational choice in alternative ways of action in an adequate manner still remains unrejected (Suppes, *Decision*, 310). The principles of rational decision theory might only be abstractly justified – though in human behavior they are often violated. Also, their potential to be up to rigid standards is limited (Levi, *The Covenant*, 1):

> Principles of rationality are ill suited for the prediction of human behavior. Nor can they be regarded as prescriptions which rational agents are obliged to obey to the letter. The reason is the same in both cases. Persons, institutions and other alleged specimens of rational agency lack the emotional or institutional stability, the memory and computational capacity, and the freedom of self-deceit required to satisfy the demands of even weak principles of coherence in belief, value and choice. Our rationality is severely bounded.

Scientists pick up problems, definitions, hypotheses, mathematical and other methods, experimental instruments, schools, research traditions, etc. One can hardly assume that – under all circumstances of choice – scientists will be familiar with all the alternatives and their possible outcomes to a sufficient extent. Choice can be conditioned by extrascientific factors of a psychological or social nature.

It will not be reasonable to expect that results of competing decisions could be evaluated with precision and could be compared as to degree of preference in all circumstances. It is an essential quality of scientific discovery that neither aims, nor means of their realization are clear-cut. This leads to the creation of new knowledge; it calls for a specific 'jump into the unknown', itself not inclusively definable by logic or calculations.

The application of the theory of decision in scientific activity calls for a deep-going, convincing philosophical justification. One can have doubts as regards its potential to reveal, by itself, the nature of scientific rationality.

Why should scientists be rational? Which one of the meanings of rationality could be applied to scientific activity? Can we reduce the requirement of coherence of beliefs of scientists as to a change in result in new empirical information? Would rationality be a characteristic feature of the whole scientific development in the world of 'the objective knowledge' (K. Popper)? The giving of answers to these questions would be the tantamount of taking a major step forward in solving the problem of the nature of scientific rationality.

In the following pages I will try to reach that goal by means of comparing, critically, two leading conceptions of scientific rationality, representative in some aspects as regards their explanatory potential: the Bayesian and the Popperian methodologies. Incidentally, one can hardly point to a theory of rationality that might, by itself, gain a complete and adequate insight into this complex matter.

2. BAYESIAN INDUCTIVISM OR POPPERIANISM

I think that one would be justified in asking the question of why – so much after K. Popper's claiming that he has solved the 'major philosophical problem' of induction (Popper, *Objective*, 1), the ambition to work out inductive logical systems continues to be unabatingly popular? Indeed, in trying to find answers to this question one should not rely on basing arguments on the dogmatism of inductivists or on the objective impossibility to find conclusive solutions of philosophical problems. Rather, it would be necessary to delve into the essence of the problem, also by means of comparing the Popperian and the inductivists' conceptions of rationality of scientific knowledge so as to outline their priorities and their drawbacks. This is a difficult research task indeed, involving reasoning on a number of different aspects. Hence, it would perhaps be reasonable to limit the task to an outline-presentation of key premises shoring up the view that the picking-out of a theory of rationality can not be exhausted by the acceptance of Bayesianism or Popperianism. Bayesian inductivism demonstrates a better philosophical and methodological fruitfulness, serious heuristic potential and logical flexibility. It is superior to Popperianism in a number of aspects; however, it does not give what is due to the intersubjective foundations of rationality. The logic of decision calls for an adequate complement, plus development in an objective – even Popperian – spirit.

1. Bayesianism upholds the prospects for pluralism in logical theories while Popperianism sets real bounds to them. According to Popper and his followers (J. Watkins, D. Miller), scientific inferences can only be deductive: starting out from true premises they would always lead to true conclusions. Basing himself on this proposition alone, one could only advance falsificationism. Popper bases his arguments on known logical incorrectness of inferences, ranging from singular

statements of the type of accounts of observational and experimental results to universal statements of the type of hypotheses and theories (Popper, *The Logic*, 27). Refutation by means of *modus ponens* is logically safe; positive confirmation of hypotheses by means of empirical testing of their deductive consequences is not so. Popper rejects quite explicitly the possibility of the existence of deductive logic: he considers as deductive the whole logic of testing and development of scientific knowledge (Popper, *On Rules*). It is beyond any doubt that this is a retreat when compared with the known Humean arguments. D. Hume also questions the logical legitimacy of inductive inferences, but he accepts their consideration in probabilistic terms. Thereby he suggests a positive solution to the problem of induction (Czezowski, *The Problem of*, 258)

It would not be justified to acknowledge the status of arguments of reasonings that are preserving the truth always and to deprive of such a privilege reasonings that are doing so only sometimes. One can not eliminate certain formal regularities, having all the characteristics of logical laws, with the exception of truth dependencies, from the sphere of logicality. That would mean neglect of the wealth of contemporary logic and would be a limitation of the possibility to have it applied to scientific thought.

The Bayesian inductive conception boils down to the application of the theory of decision in logical and epistemological analysis of science and in scientific rationality. It is a leading trend in the American and in a substantial part of the European philosophy of science. Induction is a method of construction of empirical hypotheses and theories; inductive knowledge mastering is viewed as a process of decision-making wherein – on the basis of empirical and theoretical premises – a picking-out of a hypothesis, with maximal posterior probability, is effected, in an initial set of alternatives. The aim: justification of its acceptance in the system of one's beliefs. L. Savage defines Bayesianism as a 'normative theory' seeking to be helpful in our making better decisions – by means of revealing possible contradictions in our relationships with real and hypothetical alternatives and via description of 'the changes in opinion induced by evidence on the application of Bayes' theorem (Savage, *The Writings*, 295-297). Bayesianism overcomes the neopositivist opposition of: reasoning vs. behavior, opinion vs. knowledge, value vs. inference; it perceives of logical argument as a calculation of changes in opinion, caused by new empirical information. According to works by Finnish logicians it is possible to explain the dependence on probability starting out not only from empirical data and logical premises, but also from theoretical and philosophical suppositions. Foremost representatives of the Finnish logical school (J. Hintikka, I. Niiniluoto, R. Hilpinen and others) have developed the conception of inductive systematization in which not only empirical, but also other components of the process of research – theoretical, methodological and philosophical (ontological) - play an active role. The need for theoretical terms of realization of inductive knowledge is convincingly brought out in it. The possibility for generalization of the theory of hypothetico-deductive inference through an adequate account of the inductive aspect of knowledge is forwarded by I. Niiniluoto and R. Tuomela (Niiniluoto, Tuomela, *Theoretical*). This enrichment of the logical theory of induction stems from the desire to attain conformity with the real complexity of inductive knowledge in modern science; it is, in principle, a new and progressive trend in the logic of science and the theory of rationality. There is no doubt that the fixation of a scientific fact, confirmed in experiment, presupposes inclusion, in different forms, of theoretical knowledge and methodological assumptions, as well as of overt and tacit premises, and even of ontological principles pertaining to the world picture.

2. Thus, giving due consideration to its aspects, Popperianism evidently displays an inadequate picture of knowledge that can not avoid certain contradictions and irrational presuppositions; it opposes, artificially, verification to falsification and probability to informativeness. It would seem that Bayesianism has succeeded in overcoming some of these drawbacks. Contradictions in the Popperian theory of scientific rationality account for its inadequate interpretation by I. Lakatos and T. Kuhn. They see in it certain elements of naivete, imparting an unreal weight to the procedure of falsification. Proceeding from the D-thesis, Popper would not assume the possibility for a final refutation of an established theory on the basis of *modus tollens* (Popper, *The Logic*, 50): he forwards fallibilism and critical attitude. The rationality of science finds its expression in efforts to find out critical refutation on the basis of strong logical schemes. Popper offers no rational foundations for acceptance of the empirical base of the test. Rather, he resorts to a conventional decision. Thereby he avoids, although *ad hoc*, the danger of infinite regress, but not the danger of dogmatism from in Fries' trilemma.

In a brief comment of 1968, Y. Bar-Hillel noticed that the contradiction between degree of corroboration and degree of confirmation, itself basic for the Popper-Carnap debate, is 'completely perverse'. Popper's assertion that his own criteria of comparing theories are not only different from, but are almost diametrically opposed to, those of Carnap, 'is strange' and his 'attack on inductive logic and a denial that there is such an animal cannot be explained on purely rational grounds' (Bar-Hillel, *Inductive*). There are certain facts in support of this assertion.

On the basis of one and the same formalism, inductive logicians, like Carnap, construct measures of confirmation, while the author of *The Logic of Scientific Discovery* constructs measures of corroboration. A hypothesis is better corroborated if it has held out against more vigorous efforts for its refutation. The degree of corroboration is considered as opposite to posterior probability. The first expresses severity of tests, our sincere efforts to falsify theory. The degree of corroboration is identical with its logical improbability. If there is no difference between confirmation and corroboration, from a logical point of view – then there will be a negligible difference from the point of view of philosophical methodology. Falsificationism does not conform to the real behaviour of scientists; hence it is inadequate as a descriptive thesis. It does not provide them with something different from inductive confirmation. Unsuccessful falsification does not increase the probability of a hypothesis, and it does not lead to determination of its veracity. Popper reduces the quest for truth to measuring, via abstract formulae, the degree of closeness of theories to a hypothetical, absolutely complete truth. The approbation of a most verisimilar theory presupposes falsification of all its rivals, but never a probabilistic confirmation of its truth. The conception of verisimilitude claims to be an anti-inductive account of scientific progress. However, it does not succeeded in asserting itself in this role. Inductivism and Popperianism are found to be faced by the same difficulty. If one has no right to claim that confirmation gets near to truth, then the same is also valid, to a certain extent, for corroboration or unsuccessful falsification. One can conclude that the inclusion of the concept of falsification is only a necessary addition to the inductive theory of confirmation (Wolenski, *O indukcjii*). W. Salmon is right when he says that *modus tollens* without corroboration is empty, but *modus tollens* with corroboration is inducton (Salmon, *The Justification*, 28). In fact, confirmation can only be defined by means of deductive terms – purely logical and epistemological (Stegmüller, *The Problem of*, 87).

Popper has not succeeded in avoiding compromises with psychologism. One can find only psychological justification of the expectation that scientists would strive for falsification, but not for confirmation of their theories. Quite reasonably A. Grünbaum asks the question of the meaning of Popper's statement in his deductivist conception that 'confirmations should count if they are the results of risky predictions' (Grünbaum, *Is Falsifibility*, 234). It, too, has a psychological justification. However, the great philosopher does not want to acknowledge this kind of retreat to psychologism.

On their part, Bayesians shun the psychologism and dogmatism of other inductive theories, albeit to a certain extent. Subjective probability is the expression of a degree of belief, upheld by the subject, on the grounds of the whole knowledge of, experience in and information on the truth of a sentence, or event (de Finetti, *Probability*). It could be measured through the willingness of the agent to accept a bet in favor of the truth of a hypothesis. Nobody accepts bets to his own detriment – it is a warranty for the intersubjectivity of decision (*Dutch Book)*. This requirement is identical with the coherence conditions of the system of subjective evaluations. Br. de Finnetti's Theorem of representation justifies the universal and non-psychological nature of the concept of subjective probability. It also adds the equivalence condition to coherence. This condition assumes the possibility to postulate a certain finite or infinite set of unknown objective probabilities, if true probability is not known. The representation theorem points to the connection between subjective and objective probabilities, expressed in the fact that the subjective probabilities of singular predictions of experimental results tend to the stable value of the relative frequency, therefore they can be considered as approximations to objective probability. This theorem proves that a maximization of expected utility is necessary and not a sufficient condition for rationality (Maher, *Betting on*, 30-32). L. Savage's theorem of confirmation reveals the internal structure of the process of confirmation – the knowledge of the truth of a hypothesis presupposes an ever more exact prediction of the value of its objective probability (Svetlov, *Sovremennie*, 180-188). These important results are the necessary justification in the application of Bayes's theorem to scientific reasoning, required by A. Grobler in his defence of the Popperian conception of rationality (Grobler, *Justification*, 305). He, and some other authors are right – to a certain extent – when they pay attention to the element of arbitrariness in the initial distribution of prior probabilities in the Bayesian analysis. It reflects a situation of lack of knowledge and hesitation at the beginning of every cognitive process.

Bayesianism evolves from analysis of the methodological meaning of Bayes's theorem. It relates to a set of mutually excluding, and basic knowledge exhausting, alternative hypotheses. The theorem formulates the dependence of the posterior probability of a scientific hypothesis H on the ratio of the product of its prior probability and its likelihood in the light of certain experimental evidence E, and the probability of this evidence in the light of a whole set of alternatives:

$$P(H|E) = \frac{P(H).P(E|H)}{P(E)} \qquad (1)$$

Bayes's theorem explicates the necessary connection between verification and falsification in the process of inductive knowledge: the confirmation of a hypothesis is close to 1, when the probability of its rivals tends to 0. And vice versa: its falsification in an empirical test is usually accompanied by verification of alternatives. Popper is not right in saying that the amount of compared theories always remains infinite in an empirical test (Popper, *The Logic,* 418). This is only an abstract supposition. In fact, we are faced with the case of competition, wherein the winner is the one who is considerably behind the others (R. Jeffrey). Bayesian inductive logicians are trying to avoid the difficulty of the final choice through formulation of specific rules of acceptance; some of them though, are affected by the lottery paradox (Kyburg, *Probability*, ch.14). The rules of acceptance are analogous to deductive rules of inference, but they have no cogent pragmatic justification.

The inductive rationality of science can be considered as a dynamic totality of ideals and norms, evaluative standards and rules of action (acceptance of hypotheses and theories), methods of model construction and explanation of phenomena related and applied by scientists to all scientific research. It can be understood by a pragmatized theory of inductive knowledge like Bayesianism, in its different forms and trends. The inclusion of pragmatic concepts such as acceptance, decision-making, choice, preference, and its tendency of relying on a certain universal logic of decision, testifies to a new progressive problem-shift (in the sense of I. Lakatos, in spite of his own reservations (Lakatos, *Changes*)) in the inductive program. Its main characteristic is the recovery of a new psychologism ('antipsychological' in the words of M. Notturno (Notturno, *Objectivity*)), which is in conformity with some innovatory phenomena in modern science. They are connected with increasing the theoretical attention paid to the role of the subject in scientific knowledge. It can not be an empirical individual. The subject of knowledge and reality are in a process of development and interaction. The concept of subject is an idealization – a fact guaranteeing the objectivity of logic. "The metamorphosis of science' has occurred in logic in a form of pragmatic theory of the kind of decision logic. The narrow conception of induction as an inference from the singular to the general, coming from Bacon's epoch of optimistic inductivism, cannot live up to new needs. Popper makes war against precisely this conception, thereby missing the goal.

3. On the basis of Bayesianism one can defend infallibilism and scientific realism, whereas Popperianism rejects the former in principle, as it does not succeed in justifying the latter. I. Levi considers that corpus K of our knowledge (beliefs) ought to be true. It is a standard of serious possibility and of choice of the 'best' hypothesis. If it is not true, it could not play its proper role. The functioning of this corpus of knowledge is not in conformity with fallibilism. For every accepted proposition h in this corpus we can say that 'h is true in the language L for the subject X at the moment t'.Thus the truth remains atemporal and objective (Levi, *The Enterprise*, 13-22). A scientist can put to test a hypothesis not with the aim to confirm or reject it, but to make relevant corrections to it. The choice of options that maximize the acquisition of new information with minimum risk of errors is the aim of inductive knowledge. I. Levi is right in discriminating between legitimate and illegitimate inductive expansion in his theory of inductive rationality. He even identifies both Popper and Carnap as anti-inductivists in their denying the legitimacy of inductive expansion in science. Inductive, more generally nonmonotonic, inferential relations are a generalization of deductive ones (Levi, I., *For the Sake of*, 160-193). The subject alone, and his system of true knowledge are in a process of development in the search for new information. Every cognitive act is, in itself, a transformation from ignorance to knowledge. By accepting new information, it proceeds to corrections in available information.

Bayesianism shows how subjective probabilities can approximate an objective index in results in empirical tests and can augment knowledge. The truth is a real basis for an assessment of their adequacy. According to Popper, the truth is absolutely unattainable; verisimilitude alone is a realistic goal of scientific research. However, he has not succeeded in formulating correct measures for it. Popper's conception of truth and verisimilitude is not in chime with the philosophy of active scientists. I. Niiniluoto has proved that the verisimilitude of a theory depends on the degree of its inductive justification (Niiniluoto, *What*). J. Worrall shows that the Popperian theory of rationality is feasible in practice (Worrall, *Why*).

For sure, I cannot maintain that the conclusion in favor of the Bayesian inductive conception of rationality is largely justified. It presupposes an unacceptable internalism.

3. INTERNALISM OR EXTERNALISM

Internalism presents rationality as an attribute of the thought of individual agents. For externalism, the rationality of knowledge is determined by its relation to reality and by its influence on social factors. The Bayesian decision logic is dominated by an internalist approach, while Popperianism claims to be an externalist conception of rationality.

It could not be denied that the greater part of critical arguments concerning the conceptual foundations of the inductive theory of decision are justified. Its conception of rationality is 'only formal and thin' and its account of the situations of choice is internalistic. It is subordinated to a psychological monism, which questions its relevance as a normative theory. It is reasonably observed that the judgments of subjective probabilities are 'not yet well understood' (Satz, Ferejoin, *Rational*, 72-74). The theory of decision also gives a simplified notion of deliberation by paring it down, unjustifiably, to some sort of information processing (Moya, *The Philosophy*, 94-95). M. Hesse questions the analogy between bets and scientific research (Hesse, *The Structure*, 109):

> Nature is not an ingenious opponent who seeks out just those bets, which will result in certain loss. And there is no clear relation between reasonable expectation of truth and the willingness to gamble on financial reward. The point does not need labouring that there is no close analogy between games and games against nature.

We can conclude that internalism leads the Bayesian inductive theory – to a large extent, in a wrong direction – to an inadequate explanation of scientific knowledge. Owing to this, inductivists are trying to dissociate themselves from extreme internalist subjectivism. Thus, R. Jeffrey does not consider that the settlement of opinion is the sole end of inquiry – he wants to see 'how much mileage it can get out of the fact that man is an agent' (Jeffrey, *Dracula*, 157). In another work by him he notices that probabilism, as a logic of uncertain expectations, supplements nonfoundational thinking; that there is no bedrock of certainty underlying our probabilistic judgments. Probabilism is an alternative of dogmatism, which relies on deductive logic and truth. Very often we do not have in mind judgmental probabilities for propositions, but we consider relations only, characterized by conditions for probabilities (Jeffrey, *Probability*). Fr. Schick is justified in claiming that a theory of rationality can be subjectivist without being internalist (Schick, *Making*, 136-137):

> On the theories we have presented, nothing external to you (the agent) bears on what is rational and what social. What is one or the other depends on your beliefs and desires. But this says nothing about the question of what you ought to, or might, believe and want, of what defines *oughtness* and *mightness*, here, and that is the question the internalist speaks to. Yes, we all are subjectivists...But we needn't be internalists too.

In preceding papers we have pointed to some results making up for the subjectivist orientation of Bayesianism – in its striving for an objective presentation of the development of knowledge in the process of scientific research. In this respect, one can borrow some ideas from the Popperian conception.

In an objective sense, rationality is an attribute of scientific development as a whole. Science is rational because scientific knowledge is intersubjective and overpersonal. According to Popper, the development of science is an evolutionary process going on in the World 3 of objective knowledge. His follower J. Watkins tries to harmonize the subjective and the objective approaches in the theory of rationality. If we can establish that 'certain aim is the optimum aim for science' and that we have, at hand, the one and only best corroborated theory in the respective field, then we will have 'the best possible reason to accept this theory' (Watkins, *A New View*, 73). Rationality is expressed in the successful combination of individual efforts to achieve truth plus the emphasis on objective tasks of science (problems, aims). Rationality in the subjective sense is impossible without rationality in the objective sense. We can see their relationship in the concept of norm of rationality.

4. NORMS OF SCIENTIFIC RATIONALITY

When we say that a certain action is rational, we do not mean to express an assertion, whether true or false. We express a speaker's acceptance of some norms, permitting an action. Rationality cannot be identified with justification, it can be identified with a decision to accept a norm (Maher, *Betting on*, 25). Norms are prescriptions for action and they are of an objective nature. They are in World 3. A norm regulates the behavior of particular addresses in specific circumstances by means of imposing obligations (prohibitions) and giving rights. Following J. Wolenski, we can say that norms are introduced by performative utterances of the type of: 'I state that A is obligatory (prohibited)', which expresses a decision of certain normative authority. Norm-formulation appeals to 'the will', but it is not devoid of rational grounds. A decision cannot be reduced to only its linguistic formulation; it is an act of evolvement of normative regulation, a process of forcing through an authoritative will; the result of it is a division of all possible actions into three mutually-disjoint sets: obligatory, forbidden and indifferent. In the context of logic "normation" spells out a choice of normative function (Wolenski, *Deontic*).

Normative decision is a choice called upon to substantiate the aims of normative authority. The scientific community is this authority in science. It enacts and guarantees the binding force of the norm of scientific rationality, doing it on the basis of knowledge, traditions, general view of science and its aims, and through giving an account of social factors and prescriptions. These norms are absolutely objective in the context of individual scientists. Their violation dooms a scientist's research to failure. However, this fact does not question their statute of norms. Following analogy with laws of nature, norms are independent of normative decisions. The agent makes the decision to accept rational norms or not, because such decisions fall in with his own interpretation of science's aims and problems. The choice of a norm is determined by the interest taken in maximizing epistemic utility. A norm only defines the common framework and principles of scientific problem-solving – its application is a question of professional skills and creative attitude to the particular problem. It is of no importance if the cogent's cognitive abilities do not satisfy the requirements of a norm. Such a discrepancy can be compensated by the fact that he works in a conceptual and normative framework, established by the overall scientific community. There is an invariant core in the norms of rationality, accounting for their being not so liable to change. Such a core might be the subject of another analysis.

REFERENCES

Bar-Hillel, Yehoshua. "Inductive Logic as "the" Guide of Life." *The Problem of Inductive Logic.* Lakatos, I. (ed.). North-Holland, Amsterdam: 67-69.

Bergsrom, Lars. "Some Remarks Concerning Rationality in Science." *Rationality in Science*. Hilpinen, R. (ed.). D. Reidel Publishing Co: 1-11.

Czezowski, Tadeusz. "The Problem of Induction." *Dialectics and Humanism* № 2: 257-264.

Feyerabend, Paul. *Against Method. Outline of an Anarchis Theory of* Knowledge. London, 1975.

Finetti, Bruno de. "Probability: the Subjectivistic Approach." *Contemporary Philosophy. A Survey*, V.I. Firenze, La Nuova, 1968:45-53.

Grobler, Adam. "Justification of the Empirical Basis: The Popperian vs. the Inductivist Conception of Rationality." *Analyomen* . Proceedings of the 1st Coference "Perspectives in Analytical Philosophy". Walter de Gruyter. Berlin-N.Y., 1994: 299-309.

Grunbaum, Adolf. "Is Falsifibility the Touchstone of Scientific Rationality? Karl Popper versus Inductivism." *Essays in Memory of I. Lakatos*. Cohen, R. (ed.). D. Reidel, 1976: 213-253

Hempel, Carl. *Aspects of Scientific Explanation and Other Essays*. Free Press, N.Y., 1966.

Jeffrey, Richard. *The Logic of Decision*. McGraw Hill, 1965.

Jeffrey, Richard. "Dracula Meets Wolfman: Acceptance vs. Partial Belief." *Induction, Acceptance and Rational Belief*. Swain, M. (ed.). D. Reidel, Dordrecht-Holland, 1970: 157-185.

Jeffrey, Richard."Probability and the Art of Judgment." *Observation, Experiment and Hypothesis in Modern Physical Science*. The MIT Press, 1985: 95-127.

Kyburg, Henry. *Probability and Inductive Logic*. Macmillan, 1970.

Levi, Isaak. *The Enterprise of Knowledge. An Essay on Knowledge, Credal Probability and Chance*. The MIT Press: Cambridge, Mass. and London, 1980.

Levi, Isaak. *For the Sake of Argument*. Ramsey Test Conditionals, Inductive Inference, and Nonmonotonic Reasoning. Cambridge University Press, 1996.

Levi, Isaak. *The Covenant of Reason. Rationality and Commitments of Thought*. Cambridge University Press, 1st. ed., 1997.

Hesse, Merry. *The Structure of Scientific Inference*. Oxford: Clarendon, 1976.

Howson, Colin, Peter Urbach. *Scientific Reasoning*. La Salle. Open Court, N.Y., 1989.

Lakatos, Imre. "Changes in the Problem of Inductive Logic." *The Problem of Inductive Logic*. North-Holland, Amsterdam, 1968: 315-417.

Maher, Patrick. *Betting on Theories*. N.Y., Cambridge University Press, 1993.

Moya, Carlos. *The Philosophy of Action*. Polity Press, Cambridge, 1990.

Newton-Smith, William. *The Rationality of Science*. Routledge&Kegan Paul, Boston-London-Heneley, 1981.

Niiniluoto, Ilkka, Raimo Tuomela. *Theoretical Concepts and Hypothetico-Inductive Inference*. D. Reidel, 1973.

Niiniluoto, Ilkka. "What Shall We Do with Verisimilitude." *Philosophy of Science*. Vol. 49, 1982: 181-197.

Notturno, Mark A. *Objectivity Rationality and the Third Realm: Justification and the Grounds of Psychologism: a Study of Frege and Popper*. Dordrecht, Boston, M. Nijhoff, 1985.

Popper, Karl. "On Rules of Detachment and so-called Inductive Logic. Reply to H. Kyburg." *The Problem of Inductive Logic*. Lakatos, I. (ed.). Amsterdam, North-Holland, 1968: 130-139.

Popper, Karl. *Objective Knowledge. An Evolutionary Approach*. Oxford: Clarendon Press, 1972.

Popper, Karl. *The Logic of Scientific Discovery*. Routledge, 1995.

Rorty, Richard. *Philosophy and the Mirror of Nature*. Oxford: Blakwell, 1980.

Schick, Frederic. *Making Choices. A Recasting of Decision Theory*. Cambridge, University Press, 1997.

Toulmin, Stephan. *Cosmopolis. The Hidden Agenda of Modernity*. The Free Press, N.Y., 1990.

Salmon, Weseley. "The Justification of Inductive Rules of Inference." *The Problem of Inductive Logic*. Lakatos,I. (ed.). North-Holland, Amsterdam, 1968: 22-44.

Savage, Leonard. *The Writings of Leonard Jimmie Savage*. American Statistical Association, Institut of Mathematical Statistics, 1981.

Satz, Debra, John Ferejoin. Rational Choice and Social Theory. *The Journal of Philosophy*. № 2, vol. XCI, II 1994: 71-88.

Svetlov, Vladimir. Sovremennie induktivnie koncepcii [Contemporary Inductive Conceptions], Leningrad, 1988 (in Russian).

Stegmüller, Wolfgang. "The Problem of Inductivism: Hume's Challenge and the Contemporary Answers." *Collected Pappers on Epistemology, Philosophy of Science and History of Philosophy*. Vol.II, Dordrecht: Reidel, 1977: 68-136.

Suppes, Patrick. "Decision Theory." *The Encyclopedia of Philosophy*. Edwards, P. (ed.). V. 1, 1967: 310-314.

Tuomela, Raimo. "Science, Protoscience and Pseudoscience." *Rational Changes in Science*. Pitt, J., M. Pera (eds.). Dordrecht, Reidel 1987: 83-101.

Watkins, John. "A New View of Scientific Rationality." *Rational Changes in Science*. Pitt, J., M. Perra (eds.). Dordrecht, Reidel, 1987: 63-82.

Wolenski, Jan. "Deontic Sentences, Possible Worlds and Norms. *Reports on Philosophy*. № 6, 1982: 65-73.

Wolnski, Jan. O indukcji I indukcjonizmie [On Induction and Inductivism]. W stronie logiki [On the Side of Logic]. Aureus, 1996: 284-299.

Worrall, John. "Why both Popper and Watkins Fail to Solve Problem of Induction." *Freedom and Rationality. Essays in honour of John Watkins*. D'Agostino, Fr.(ed.). Kluwer, 1989: 257-296.

SPAS SPASSOV

THE INFORMATION TECHNOLOGY REVOLUTION: A NEW TECHNO-ECONOMIC PARADIGM

1. WAVES, REVOLUTIONS, PARADIGMS: A NOTE ON METHODOLOGY

Every analysis of a phenomenon, be it of a scientific, economic, social, or another character, should start with an analytical effort to make the fundamental concepts and ideas involved as clear as possible. The phenomenon of Information Technology Revolution has been treated in the literature from different points of view and characterised by different terms, stressing one or another aspect of its multifaceted nature or manifestations. Although there is much common understanding of what the main characteristics of this phenomenon are, different emphases associated with different terms may lead to unnecessary misunderstandings. It is important therefore to outline the main ideas and concepts that will underlie this essay.

Daniel Bell speaks of three technological revolutions occurring in modern times in the Western world, the IT revolution being the third one. The first revolution is associated with the introduction of steam power by the end of the XVIII century, which provided a significant increase in people's ability to apply energy to machines. The second one occurred about a hundred years later, and is identified with two innovations: electricity and chemistry, one of them giving a new, enhanced form of energy that can be transmitted easily to a long distance thus permitting new kinds of decentralization, and the other allowing one to create new materials unknown in nature (Bell, 1989, 164-165). This view seems to be widely shared and to reflect a common, though quite general picture of technological revolutions. What is important in Bell's account, is the explicit emphasis on the distinction between a technological revolution and its socio-economic consequences. The distinction points out the fact that there is no teleology in these processes of technological innovation, i.e., that no single necessary and determinate path to the use of the new technologies exists. The fact that there is a great diversity in the ways the new technologies can be organized refers to the central role of the social and economic dimensions of these revolutions.

A more elaborate analysis of technological innovations in the modern history of Western society is provided by the authors who follow the approach of Kondratieff and Schumpeter. This approach describes the waves or cycles of technological

Dimitri Ginev (ed.), Bulgarian Studies in the Philosophy of Science, 31-41.

change leading to successive industrial revolutions. The essential feature of these industrial revolutions points to a qualitative transformation of the way the whole economy is organized by the new technologies. Authors following this approach differ in the way they explain the long waves, but the important idea they share points to the necessity to analyse the fundamental changes that affect a wide range of industries and have long reaching repercussions on society and culture, rather than the growth or transformation of individual industries. Freeman and Soete distinguish among five such waves, starting with the industrial revolution, which occurred between 1780s and 1840s, and each of them lasting nearly about 60 years. The last wave started in the 1990s and is characterised by the introduction of microelectronics and computer networks. The central idea that underlies their explanation of these waves is based on the central role of the widespread diffusion of numerous innovations based on a new infrastructure (Freeman and Soete, 1997, 20).

The explanation of the long-term impact of new technologies on the economy requires a distinction between differing levels of innovations. For the more general purpose of describing the technological revolutions it is sufficient to distinguish between two such levels: incremental and radical innovations. Incremental innovations occur as regular improvements in various economic branches and may have a significant impact on the overall economic growth and productivity, but their economic effect is limited and they do not lead to the introduction of a new mix of products and services affecting the whole economy. In contrast, radical innovations result in the introduction of unique products and processes with a broader effect on the overall economic performance. This effect may take the form of significant structural changes in the economy as a result of the emergence of new industrial branches, or the opening of new markets for the new products, new capital equipment, new type of skills required and so on (Wolfe, 4-5).

One distinctive feature of radical innovations is that they often occur in clusters, which some authors call constellations of innovations that are linked to each other in both their technological and economic characteristics. Such constellations of innovations affect several branches of the economy and may lead to the emergence of new branches. When a profound transformation of the conditions of production and distribution occurs in potentially all economic sectors, based on a constellation of related innovations, one can speak of a technological revolution. Some authors associate the technological revolutions with new techno-economic paradigms. The main features of the new techno-economic paradigm are a rapidly falling cost curve related to the core new technology, a virtually unlimited source of supply over a long period of time, and a potential for the adoption of this technology in a wide range of industries and branches of the economy. The emergence of a new techno-economic paradigm has a significant disruptive effect on the economy as a whole, which requires a long-term structural adjustment (Wolfe, 5-6).

It is not one of the objectives of this paper to pursue a more elaborate analysis of the literature on this question. What emerges from this overview is the outline of an understanding of the phenomenon of technological revolution, which will guide the analysis of the technological and other changes related to the present day introduction of information technologies. The terms 'technological revolution' and 'techno-economic paradigm' will be used interchangeably[1], although the second one better reflects the idea of the complex character of this phenomenon and the central role of the organizational socio-economic and cultural transformations associated with the new core technology. A technological revolution implies long-term radical transformations in many characteristics of social and economic life and affects the way society is organized. As Manuel Castells puts it, technological revolutions are inducing a pattern of discontinuity in the material basis of economy, society, and culture:

> ... they are all characterized by their *pervasiveness*, that is by their penetration of all domains of human activity, not as an exogenous source of impact, but as the fabric in which such activity is woven. In other words, *they are process-oriented*, besides inducing new products (Castells, 1996, 30-31).

All technological revolutions imply three common features: 1) accessibility, i.e., availability in virtually unlimited supply; 2) dramatically declining price curve; 3) potential applicability in a large range of industrial branches and economic sectors. These features will be examined with regard to the new techno-economic paradigm that is shaping our lives in the present era.

2. A HISTORICAL SURVEY OF THE INFORMATION TECHNOLOGY REVOLUTION

The precise historical sequence of the information technology revolution, including its starting point and the key elements in its development, vary in different accounts, which reflects the way different authors view the underlying processes and interpret the overall emergence of the new techno-economic paradigm. As was mentioned above, according to Freeman and Soete the last wave of technological change started during the 1990s and is developing through the present time. In order to analyse the accompanying social, economic and political restructuring that is equally important, it is necessary to take into account some other relevant processes that took place earlier. Many authors point to the 1970s as the starting point of the new techno-economic paradigm, with some key technological innovations occurring in the period immediately after the second world war. This view is more appropriate for the understanding of the large-scale of the changes that started during this period and that are continuing to the present, still shaping our present and our immediate future as a society.

According to Daniel Bell, four significant technological innovations underlie the new technological revolution. The first consists of the switch from mechanical, electric, and electromechanical systems to electronics. This process resulted in a significant reduction in the number of parts, and in a dramatic increase in the speed of transmission. The second innovation is linked to the miniaturisation of the elements conducting electricity or switching electrical impulses. The main result of this process is a significant increase in the range of potential applications of the new technology. The third innovation is digitalisation of information. Through the use of digital switches and binary systems, all previous systems were converted from analogue into digital form. Finally, the introduction of easily accessible software provides the basis of the customisation of the new technology, thus opening the road to the widespread use of personal computers in small businesses and private lives (Bell, 1989, 165-166).

Manuel Castells traces the history of the new electronic-based technologies in three main technological fields: microelectronics, computers and telecommunications. He considers as a main starting point the invention of the transistor in 1947, which provided a means for faster processing of impulses in a binary mode. The widespread use of semiconductors became possible with the introduction of new manufacturing technologies and the switch to silicon as a building material, which made possible miniaturisation and the rapid decrease in the cost of production. The next decisive step in the development of microelectronics occurred in 1957, with the invention of the integrated circuit. This invention led to a dramatic fall in the prices of semiconductors (by 85% between 1959 and 1962), and to an increase of production by 20 times for the next ten years. These processes accelerated during the 1960s (Castells, 1996, 41-42).

The widespread diffusion of the new information technologies that led to the emergence of the new techno-economic paradigm, took place largely during the 1970s. One of the main factors contributing to the overall impact of the new technologies during this period is the rapid miniaturisation of processing units and the accompanying spectacular decrease in the cost of production. Many authors consider this period as crucial for the beginning of the information technology revolution. It is important to take a closer look at the main innovations that occurred during this time, though avoiding any technical details.

The main effects of the introduction of low-powered computing devices using semiconductors are twofold: they provided a potential for significant gain in speed, thus outperforming the big macrocomputers of the previous generation, which used vacuum tubes and had limits to their increase in power, and they allowed for a dramatic decrease in the cost of production, thus opening the road to the rise of the personal computer industry. The combination of these two effects was crucial for the widespread adoption of information technology in virtually all sectors of the economy. Three key innovations led to the introduction of microcomputers: the working memory, the software memory, and the microprocessor CPU, built entirely into single chip. All these innovations were made by Intel between 1969 and 1971. Although the use of integrated circuits instead of transistors in the computer CPUs started earlier, these machines had to use hundreds of circuits and needed electromechanical memories and peripherals. The engineers from Intel reduced the number of chips to three: one for storing the mathematical programmes and operating systems (the read-only memory), one for the working memory (RAM), and one for the CPU, which represented an entire computer on a single chip. These three innovations perfectly complement each other, as without the microprocessor, the silicon memories represent nothing but a minor improvement on the previous macrocomputers, and if not linked to these memories the microprocessor would have been only a new form of integrated circuit with a limited scope of performance. The combination of these three innovations made possible the significant jump in efficiency that ultimately led to the widespread use of microcomputers.

It is interesting to note that the enormous potential for a widespread application of this new threefold invention was not realised for some time by both Intel and the other players in the industry. The Intel designers even closed the microprocessor project and turned their attention to other fields of research. It was the unexpected market success of the microchip that pushed them to improve its design and to release new versions of it, which eventually led to the rise of the microcomputer industry. The application of the microprocessor in several unexpected areas created a huge demand for the new device, which turned the purely technical invention into a widespread technological innovation. The crucial role played by market forces in this case is only a curious illustration of the whole phenomenon of the information technology revolution as a market-driven process. The introduction of microcomputers and the miniaturisation of technology led to smaller, faster, and cheaper products. This is what made possible the enormous potential for application in virtually all sectors of the economy, including the introduction of personal computers. Thus, the new direction of technological innovation toward microtechnology eventually translated into widespread application, efficiency and profitability, which in turn shaped the microtechnological revolution.

Another important phenomenon, which occurred almost at the same time as the introduction of microelectronic technology, is the revolution in telecommunications technology. The first significant innovation in this field was the replacement of electro-mechanical switches by electronic ones, which had started by the mid 1970s. The digital switch offered enormous advantages compared to the analogue one in terms of increasing speed, power, flexibility, energy and labour efficiency, and low cost (Castells, 1996, 45-46). The possibility of transmitting information, including data, voice, and later on images, in a digital manner, combined with the widespread use of personal computers storing the same information in digital form, led to a growing convergence of computer technologies and telecommunications technologies. This convergence of information technologies continued to develop in the 1980s and 1990s with the introduction of multimedia computing (Wolfe, 10).

These significant innovations in the microelectronic and telecommunications technologies made possible the next important step in the formation of the new techno-economic paradigm: networking. The initiative that led to the creation of what later became known as the Internet started in the late 1960s as an effort of the US Department of Defense to create a new communications network, able to transmit information and provide access to other computers within a decentralised system of military sites. The development of the interconnection protocol TCP/IP in the mid 1970s allowed the integration of the US network with other similar networks in different countries, thus growing into a global Internet, which eventually included a large number of rapidly created local area networks at universities, government agencies and large corporations (Wolfe, 11-12). The broad public usage of the Internet, though, became possible after the creation of the World Wide Web, which was based on the combination of an addressing system for local information on the Internet and a means of linking this information using the hypertext bookmark device. The subsequent creation of the Web browser, which allowed any user to easily navigate the web, led to a spectacular increase in the number of web sites and interconnected computers (*ibid.*, 12).

It is worth noting the important role played by the military sector in the emergence and shaping of the information technology revolution. The role of the military was twofold: as a significant centre for research and development and provider of new technologies and innovations, as the example of the Internet illustrates, and as a major consumer of new technologies, thus intervening in a significant way on the demand side of the relationship. For example, during the period of rapid increase in the use of semiconductors in the US in the 1960s, 50% of it went to military uses (Castells, 1996, 42). Similar figures are likely to be found in the present day economy, given the size of the US military budget and its emphasis on new technologies. These facts may be interpreted in support of some historical theories, which attribute to military endeavours throughout history a paramount role in the advancement of science and technology. It is not possible here to speculate on this subject, but it is nevertheless important to stress the significant role of the military sector, and through it of the state, in the processes of emergence and development of the new techno-economic paradigm.

3. THE MAIN FEATURES OF THE INFORMATION TECHNOLOGY REVOLUTION

The historical account of the main technological innovations that underlie the rise of the information technology revolution points to the 1970s as a crucial period and indeed the true beginning of this process. This fact is stressed by many authors, and the attempt to explain the reasons behind it can help one to understand some of the important features of the new techno-economic paradigm. As mentioned before, the first break-throughs in the electronic technology that led to the construction of electronic computers were in place much earlier, and during the 1940s and 1950s there were already forecasts about the many potential applications of this technology throughout the economic system. The question of why it took a few decades until the widespread use of electronic technology became a fact, is not an easy one to answer. There are, of course, many different reasons that interfere with each other and form the complex process that shaped the path towards the new techno-economic paradigm. Some of these reasons are historical and accidental, others follow from the internal logic of the process of technological innovation itself. On the historical side, the most important factor is the beginning of an overall economic crisis in the industrialised world by the early 1970s. The period of rapid economic accumulation after the second world war, known as the Golden Age of contemporary capitalism, reached its limits and exhausted its potential for further growth. This created the conditions and the necessity of a profound restructuring of the whole socio-economic paradigm established in the post-war period. The introduction of the new information technology and its widespread use throughout the economy coincided with these macroeconomic processes, which is certainly not a mere historical coincidence. The emergence of the new information technology played a twofold role on this level: it deepened the economic and social crisis that had already started by that period, and at the same time it became an essential part of the restructuring process that affected socio-economic organization and shaped the way society functions for the following decades.

The comparative study of periods of technological revolutions in the modern world reveals some common features and trends that point to an internal logic of these processes. Although it seems that the literature is still far from providing a full description and understanding of this internal logic, some important features clearly emerge. In all cases, the establishment of the new techno-economic paradigm always takes a prolonged period of time, because in addition to a cluster of technical innovations it requires and involves a wide range of significant educational, social, and managerial changes. These changes are accompanied by the development of new social institutions that have to respond to the needs of the new technology. The role of social institutions may be in opposition. The conservatism and inflexibility of the existing institutions may slow down the diffusion of the new technology, while the creation of new institutions may considerably facilitate and accelerate this process (Freeman and Soete, 1997, 184-185).

The necessary conditions for the establishment of the new information technology paradigm started to develop during the 1950s, 1960s and 1970s. At that time, a massive capital goods supply industry was built on a sufficient scale to provide the necessary equipment for the technological transformation. The electronic technology provided enormous improvement and advantages in terms of relative costs and reliability compared to the previous electro-mechanical systems. At the same time, the necessity to train sufficient numbers of people able to design, operate and maintain the various processes involved in the new technology became one of the main obstacles to its widespread diffusion, because it obviously needed a long period of time. The skill shortage was particularly acute in those sectors of the economy that had previously been outside the scope of the electronics technology. Another obstacle to the rapid diffusion of the new technology was the high cost of software design, which remained relatively high even when the cost of hardware dramatically dropped. In addition, in many service sectors the adoption of information technology was possible only after some legislative, organizational, and managerial changes were brought about, which required a relatively long time (Freeman and Soete, 1997, 185). As a result of that, the new information technology was first widely adopted in those economic sectors that experienced a rapid growth and attracted a lot of new investments. It is not surprising that initially, microelectronic technology found widespread application in the electronics industry itself, which was the single most rapidly expanding sector of the economy.

The involvement of a wide range of transformations that are external to the technological innovation itself can explain the relatively slow pace of adoption of the new technology, especially in those old sectors of the economy where the growth was slower and the number of people qualified to deal with this technology was lower. According to some authors, the most important new technologies usually start in the public domain in government laboratories and universities (with an important share for the military sector), then move to instrument sectors and large manufacturers with strong R&D, and finally go to the large consumer goods industries and services (Freeman and Soete, 1997, 186). These trends point out once again the crucial role of the state in the shaping and the pace of the establishment of the new techno-economic paradigm.

4. THE SOCIAL IMPACT OF THE INFORMATION TECHNOLOGY REVOLUTION

One of the most important social effects of the information technology revolution is related to the significant changes it brought about in the nature of work and the structure of employment. The new technology and the restructuring of the whole economy that it provoked are affecting directly or indirectly the nature and structure of the work force at all levels. These changes, which started in the 1970s and are going on in the present time, can be compared to the similar transformation in the work force during the industrial revolution in the beginning of the century. According to Castells, during the industrial revolution societies became post-agricultural, while during the information technology revolution they became post-industrial (Castells, 1996, 208). In contemporary economies, work is increasingly computer mediated, which means less physical and more reliant on microelectronic technology. This trend leads to increasing segregation of production from design and conception, and to a dominant role of the processes of control and design over production.

The most visible change in the structure of employment taking place in the present technological revolution, is the rapid decline of manufacturing employment as a result of the overall process of economic restructuring and technological transformation. While this trend has been observed in all developed countries, there are significant differences from country to country, which points to the important role that different socio-economic conditions and government economic policies play in the shaping of the way employment is affected by technological transformation. It is extremely difficult, however, to create a more accurate and detailed picture of the complex processes of employment restructuring within different sectors of the economy or on the company level. Difficulties are both methodological and empirical. As some studies show, applying different methodologies to one phenomenon under consideration often leads to quite divergent results. In addition, there are many other factors that occur at the same time as technological innovation that also affect the structure of employment. The most important such factor is international trade, which may lead to job losses in certain industries, although it is considered that its impact on aggregate employment is negligible. The impact of international trade and that of technology innovation on employment structure cannot be examined separately, because today's international trade is increasingly technology driven and the effect of technology on jobs is to a large extent determined in international markets (OECD Report, 75). Nevertheless, the available evidence can help one to understand the main trends in the restructuring of employment due to the information technology revolution, and the crucial role of government policies and company strategies in these processes.

On the level of the global economy, there are two main trends in the developed countries during the last thirty years: the progressively weaker overall employment growth and the shift from industries to services (OECD Report, 52-53). The overall annual employment growth fell from about 2 per cent in the 1960s to 0.5 per cent in the 1990s. In the services sector, however, the job expansion remained stable at over 2 per cent per year, while in industry it was very weak or even declining. The differences in these trends between the countries and the regions are considerable. The combined job creation in services and industry in Europe was much slower than in North America and Japan. In North America, the services sector was very dynamic, and jobs in industry as a whole increased slightly, while in manufacturing they declined sharply in the 1980s. In the EU, the growth of employment in the services sector remained slow and steady, while jobs in industry declined and by the 1990s were being lost at the rate of 3 per cent a year. In contrast, in Japan employment growth in the industry as a whole remained positive and steady during almost the whole period. Since the early 1990s, jobs in manufacturing have declined sharply and those in the services sector have risen at a fast pace (*ibid.*, 54). These differences clearly illustrate the crucial role of different government policies in the way the technological transformation affects the structure of employment, although a more concrete and detailed picture would require an elaborate economic analysis, which is not possible in this paper.

On the level of individual firms, the impact of technological innovation on employment is difficult to be assess, but some empirical studies show that as a general rule technologically advanced firms have a higher rate of employment performance, while low productivity firms in the same industry shrink through downsizing. The adoption of new labour-saving technologies tends to decrease employment as a whole, although in some innovating firms employment may increase. This trend is particularly clear in firms engaged in product innovations, where positive employment growth is a result of the increased demand for the new product, which in turn generates higher demand for labour (*ibid.*, 62-63). It can be concluded that high productivity growth due to the introduction of new technologies does not necessarily translate into job losses, because it may be accompanied by other factors, such as higher incomes and increased demand, which may lead to employment growth. This fact is important, as the evidence shows that most of the structural changes in employment due to the technological innovation occur within the industries, on the firm level, rather than in the industry as a whole, where the rate of net job change is relatively low.

The positive impact of technological innovation on employment growth can be observed more clearly in the services sector, although there are many other factors besides technology that affect jobs in this area. The direct impact of technological innovation on employment can lead both to increase and to decrease of employment. The introduction of new processes and new service products can displace employees and result in job losses, but it can also lead to job gains by creating higher demand and hence increasing employment. Indirectly, technological innovation affects services jobs by raising productivity and thus leading to increases in wages and incomes. Since in the developed countries increasing part of incomes is spent on services, these trends lead to increasing an demand for services jobs. According to some studies, although the introduction of new technologies in the services sector increases productivity and leads to restructuring in employment similar to that in the industry sector, new demand and new jobs creation outperform the job losses. In countries that invested more in the application of new technologies, employment gains tend to be larger (*ibid.*, 73-75).

In the services sector, as well as in manufacturing, technological innovation leads as a whole to a decline of low-technology and low-skill jobs, and to an increase in high-skill and high-wage employment. Those industries and firms that have fastest productivity growth also have higher job gains. The introduction of new technologies in general tends to increase real incomes through higher wages and profits or through price reduction, which creates increasing demand for goods, especially new products, and services, thus leading to employment growth. For example, according to some studies, high-technology industries in the G-7 countries have had above-average productivity gains and at the same time better than average employment performance, while in low-technology industries both productivity growth and employment performance have been below average (*ibid.*, p. 66). It is important, however, to note that technological innovation does not necessarily lead to higher employment performance. The evidence on the firm-level suggests that higher productivity and employment can only be achieved by complementing the technological innovations with organizational changes (*ibid.*, 77).

5. CONCLUSION

The main purpose of all efforts to understand the nature and internal logic of the processes associated with the establishment of the new techno-economic paradigm, its characterising features, and its impact on society, is to construct a better picture of the range of possible choices that societies have, and ultimately to formulate well informed social policies in response to these processes. One of the most important ideas that emerges from the studies on the information technology revolution refers to the fact that there is no one single necessary and determinate path for the introduction of the new technologies. There is a variety of ways the technologies can be organized and used, and which of them will prevail depends to a large extent on the social choices and decisions. In other words, the establishment of the new techno-economic paradigm is not a technology driven process, technological determinism is not the main factor that shapes the path of the technological innovation. As D. Bell puts it,

> Technology does not determine social change; technology provides instrumentalities and potentialities. The ways that these are used are social choices (Bell. 1989, 167).

In this context, it is obvious that the role of the state not only in shaping the path of the information technology revolution, but also in the attempt to put under control its social impact, is crucial. As the analysis shows, economic policies and social choices may significantly affect the way technological innovation affects the overall process of socio-economic restructuring in the developed world (which accounts to a large extent for the differences between countries and regions). The role of the state should not be narrowly understood as a direct intervention on the socio-economic scene. The experience of the last thirty years points to the fact that the trend towards increasing decentralization and delegation of government functions to the private sector and NGOs better matches the very nature of the information technology revolution. It can be argued, for example, that the decision of the US government in the 1980s to deregulate telecommunications played a crucial role in the explosion of the Internet network and the World Wide Web in the mid 1990s. As the WWW is the most important event in the present day technological revolution, this decision has had very important consequences for the socio-economic restructuring that will shape the future in the short and middle term. However, it was a rational choice made by a government that has to assume a huge responsibility for such decisions. In the long term, the way resources are allocated now will determine technological and economic development in the future. As technological knowledge is the key element in the process of technological innovation, present educational policies are one of the main instruments for societies to respond to the challenges of the new techno-economic paradigm. It is by making thoughtful and well informed rational choices that the state can and should play a central role in the shaping of the information technology revolution.

NOTES

1. It seems that authors writing of technological revolutions make explicit or implicit reference to T. Kuhn's famous account of scientific revolutions. The analogy however is limited and should not be pushed too far. Scientific and techno-economic revolutions are very different phenomena in their natures and their structures. Thus, many fundamental concepts in Kuhn's theory (e.g., normal science, scientific crisis, scientific community and so on) are simply irrelevant to the technological revolution. This theory, therefore, cannot be used as a model or a methodological guide for the analysis of the techno-economic transformations. Still, since the terms 'revolution' and 'paradigm' had been legitimately used long before Kuhn's revolutionary book on scientific transformations, they are very well suited for economic and technological analysis, without necessarily making reference to Kuhn's work.

REFERENCES

Arnold, E. and Guy, K. "Policy Options for Promoting Growth Through Information Technology." OECD. *Information Technology and New Growth Opportunities*, 1998.
Bell, D. "The Third Industrial Revolution." *Dissent*. Spring 1989.
Castells, Manuel. *The Rise of the Network Society*. Oxford: Blackwell Publishers, 1996.
Glyn, A. et al. "The Rise and Fall of the Golden Age." *The Golden Age of Capitalism: Reinterpreting the Postwar Experience*. Stephen Marglin and Juliet Schor (eds.). Oxford: Clarendon Press, 1990.
Freeman, C. and Soete, L. *The Economics of Industrial Innovation*. Third Edition. London: Pinter, 1997.
"The OECD Jobs Strategy." *Technology, Productivity and Job Creation*. Vol. 2: Analytical Report, 1995.
Wolfe, D. "The Rise of a New Techno-Economic Paradigm." (unpublished).

ASSEN PETROV

ARE BIFURCATIONS OF HUMAN KNOWLEDGE POSSIBLE?

This text emerged from fragmentary reflections about physics, science and language. It is an attempt to organize the key ideas around a test question that seems to pervade, though implicitly, the present and future days of the philosophy of science. "Human knowledge" is something much bigger than science proper, but it is science where the question is highly non-trivial. In this perspective I focus on science.

1. TWO BASE CAMPS

The major issue is whether the existence of competing scientific ideas can result in irreversible splittings of a science (or the science) into equally viable and no-longer-interacting mental constructs. No competition, no argument, no common language. That is what we call a genuine bifurcation of scientific knowledge. Are such bifurcations possible? How far could the paths of a split science diverge? Are there noticeable bifurcation trends? If so, should we resist or should we foster them?

Two extreme answers are known – a categorical "no" and no less categorical "yes".

The naïve position of traditional rationality rules out any thought of bifurcation on the authority of the physical reality as the ultimate factor determining the totality of natural science. Our knowledge may be riddled with gross errors but the external world is a reliable partner, an entity unique, unchangeable and knowable. The scientific ideas and pictures of the world can only be involved in temporary rivalry, they are bound to converge over the course of time. A simple, clear-cut position. The traditionalist could at most have some doubts whether we could ever reach the complete objective truth about the physical world.

In contrast to all this, the extreme liberalism rejects the dictates of the external world as fictitious, useless prejudices, the concepts of external world and objective truth themselves are only obstacles to human creativity. We are free to call into being and sustain arbitrary scientific ideas in pursuit of our life goals. Cumulativeness and convergence are delusions. The difference between a scientific treatise and a fictional text is negligible, scientific pluralism is not only possible, it is desirable and unavoidable. Not truth, but free, limitless creativity is a supreme value. An appealing realm of freedom, no doubt. What the extreme liberal should

Dimitri Ginev (ed.), Bulgarian Studies in the Philosophy of Science, 43-49.

worry about is just the observation that present-day science is intolerant to continued ambivalence and it maintains a good relationship with the external world.

How are we supposed to behave in the presence of opposing schools of thought, each of them underpinned by weighty human values? In this particular case we are going to look for a common context and even common values in order to preserve and harmonize what is essential in the traditional and liberal approaches.

2. THE POINT OF COMPROMISE: COHERENCE

If we survey the development of science, we cannot miss the constantly increasing subjective component in the description of the objective world. Rising autonomy in shaping the world's image means, of course, growing epistemological responsibility for the knowing subject and undermining the traditionalist's way of reasoning. Let us look into the compromises that are to be expected.

The traditionalists have to make the first, crucial moves. Their line of argument can be summarized in the implication chain:

external world => convergence => coherence => uniqueness.

Much of what follows is related to this chain.

Unlike the rigidity of the external world (it is there or it is not), the coherence is flexible and elusive. We do not know what "complete coherence" can mean, yet we feel that there is a lack of coherence everywhere in science (and no coherence vacuum at the same time). Let us venture a provisional definition. From a somewhat pragmatic point of view we can say that a collection of ideas (concepts, statements, theories) is coherent if the removal of any part of it affects the functioning of the collection as a whole. Outside science a coherent collection of ideas may consist of unrelated or even contradictory or meaningless elements, if its intended function requires it. Coherence, therefore has little to do (even in science) with logic. Truth appears now as a measure of the level of coherence. This coherence-based concept of truth is akin to the consensus truth, though a closer study would show some differences.

The coherence increase is not a side-effect of the development of science. Much more the search for coherence is a major driving force, a high degree and a great scope of coherence being first-rank scientific values. A serious lack of coherence is a potential threat of grave splits in science.

The coherence is important for our purposes because it is the last (and therefore the most reliable) element in the chain preceding the uniqueness, directly underlying our belief that all splits in science are transient phenomena. It marks the limit of the compromises the traditionalists can make, if they see themselves forced to abandon the notion of the external world as a rock-solid guarantee generating the whole chain. Indeed, the role of the external world shrinks and the convergence becomes questionable as the subject's autonomy expands, while the pursuit of coherence remains very characteristic everywhere in science and seems capable of overriding all trends to non-uniqueness. The fact that science now presents an embarrassingly incoherent picture and the suspicion that complete coherence (whatever this can mean) is forever out of reach are irrelevant – what could stop us from removing the incoherences we encounter one by one and thus counter the threats of serious splits?

Let us stress this: in their retreat the traditionalists do not discard the notions of external world, physical reality and all other attributes in the common sense. They only deprive the external reality of its absolute outside-ness, insisting on the right (or rather taking the burden of the responsibility) to control its place and role in the world of human ideas.

Yet the concessions made are significant. If we drop the first two terms of the implication chain, the development of science becomes essentially a random process. Unlike the external world or the ultimate convergence, the knowing subject's control over the changes in science can be only local. On a sufficiently large time-scale, it is chance that decides which incoherences would determine the next turn of science (which can be nothing less than a "revolution"). The path science follows can be likened to a series of what in mathematics is known as "random walks".

This randomness is something we have to come to terms with. When cumulativeness and coherence clash, the cumulativeness falls apart, the science may change its course sharply, but no lasting splits occur. Coherence is always victorious – that is the characteristic of what has been happening in science for centuries, up to the present day. (By the way, this means that science is inherently cumulative, if we monitor its coherence, not its content.) As long as the coherence is the winner, the uniqueness of science is safe, which is the way the traditionalists can argue. Science making random walks is something unpleasant, but the real threat –that of bifurcation – is kept at bay.

One can almost see the liberal's ironic smile at the word "threat". Indeed, there is an elegant irony here, but it lies somewhat deeper: the coherence would eventually undermine its own role as a safeguard against grave splittings. Growing coherence means bigger and more rigid structures in science and therewith the cost of a scientific revolution rises steeply – it is not the same if we have to revise our notions about the free fall of bodies on the earth's surface, or the theory describing the evolution of the universe. At some point in the future the price will become too high, and we cannot be wrong if we predict that science will some day face major dilemmas of the kind coherence vs. stability that are critically difficult to decide. The moments when the choice is made in favor of stability and against coherence are milestones in the life of science. They amount to deliberately postponed revolutions, a crucial step towards legalizing the idea that science can sustain bifurcations and remain science.

A triumph for the liberal? Not quite. There is additional irony in the fact that the way for the irreversible splits in science is paved by such traditional conservative values as coherence, cumulativeness, stability and continuity.

Bifurcations are very likely unavoidable, we already glimpse them in the far future. We now recognize the full weight of the epistemological responsibility, as we must learn to formulate and answer questions about the precarious balance between coherence and stability, preparing ourselves for dealing with crossroads dilemmas. Briefly, we have to obtain insight into the anti-split mechanism and its neutralization at critical points. Matters about which we know practically nothing.

3. THE BIG PICTURE: SCIENCE IN THE WORLD OF IDEAS

Our observations and conclusions from Sec. 2 appear more natural and convincing, if we think of science as embedded into the all-encompassing world of human ideas, where it actually belongs. The notions we used, coherence, stability, cumulativeness, anti-split mechanism etc, all make sense when applied to the humanities, mathematics, religion, art, history, economics and philosophy. Outside science the pluralism of ideas is their natural state. We do not encounter distinctly articulated dilemmas of the sort coherence – stability, but an anti-split mechanism is still at work as there is no trend towards infinite divergence and disjunctive bifurcations.

The great advantage of the ideas (or of language, or of texts – we do not need to distinguish between them here) is that they form a homogeneous space, the ideas (words, texts) are freely interchangeable, and any two of them can be identified, if we decide so. Fascinated by this homogeneity, one can easily yield to the temptation to ignore the internal architecture in the world of ideas, delete all boundaries and time-honored distinctions (subject – object, science – humanities, empirical – theoretical, theory – criticism ...) and deny the privileged role of some concepts or hierarchies (consciousness, science – philosophy). Being aware of such limitless freedom is indeed an inspiring sensation and offers inexhaustible room for manoeuver. The extreme liberal's attitude to science stems from such intoxication with freedom.

Sobering remark: the structureless homogeneous space of ideas is only an empty stage where non-trivial things happen (in science and everywhere else). We cannot stop and contemplate forever the homogeneous desert. Everything interesting or just meaningful is necessarily a disturbance (always a transient one) of the homogeneous language continuum.

The rise and fall of privileges in the space of ideas and the restructuring of its architecture belong to human cultural history (not to be separated from general history), they cannot be described as philosophical revelations. The space of ideas has not always existed as a universal homogeneous space. The same can be said about its rival space, the external physical world. The words "idea" and "external physical thing" are homogenizers giving rise to two homogeneous spaces, each one pretending to contain the other and yet clearly different... The history (and archaeology) of ideas can say something (very little) about how previous homogenizers (mythologies, religions ...) are gradually and slowly replaced by "idea" and "external physical thing" and how quite recently "language" has taken the privileged place of a superhomogenizer.

The leading thread through all these developments (reaching far back into prehistory) is the constant increase in the subject's autonomy and control over its mental activities. The steadily growing epistemological responsibility, stressed in Sec. 2, fits perfectly into this global picture. We are then entitled to assume that the knowing subject would end up "committing suicide" after having swallowed and digested the external world. (The subject/object opposition and the mind/body problem are marginalized and can disappear before any other solution has been found.)

Another scenario is that the external world be preserved as a useful simulation, a scaffolding for the needs of the anti-split mechanism. In any case: the presumable demise of the knowing subject does not imply the end of the subject as a privileged element in the space of ideas – the conscious creature, thinking and speaking, the bearer of the space of ideas, persists.

Are we not moving towards the end of the science? Perhaps, but not towards a "final theory" or a "theory of everything". The scenario leads to something more serious – science without object, scientific language that does not refer to anything outside itself.

Science needs then new means to redefine its identity and defend its legitimacy, different from simply referring to the external world or objective truth. Anticipating the imminent identity crisis, the first thing one could do is to assume that "science" denotes that area of ideas where all splits are only signs of growth, never allowed to develop into permanent bifurcations. This identity card corresponds to the present state of affairs and is not likely to survive the serious crisis lying ahead. We prefer a much more relaxed definition: science is characterized by the (relatively) highest degree of conscious control over the immanent large-scale randomness of events in the space of ideas. Equivalently: science means the highest degree of epistemological (or rather intellectual) responsibility. The transition from the first to the second definition is actually the crisis itself; we are trying here to guess its outcome.

The image of science we delineated is complex, almost intimidating: debunking of absolutes, clashing human values, constant restructuring of the space of ideas, new homogenizers, sequence of random walks, context-dependent coherence-based truth, delicate balance between coherence and stability, redoubtable tasks related to the work of the anti-split mechanism, critical dilemmas looming ahead, struggle to save the identity of science. We tried to interpolate between the naïve rationalism and the extreme liberalism and ended up with this down-to-earth image. It is summed up by the notion of epistemological responsibility. Clearly, the traditionalist and the liberal choose opposite directions to avoid this responsibility.

How would the two camps respond to the suggested compromise? Would the liberals ever accept the full burden of epistemological responsibility as a worthwhile substitute for their utopian freeplay of ideas? Would the traditionalists ever recognize the human autonomy in science as a value and accept the epistemological responsibility as a reasonable price to pay for it? It is hard to say: so far the traditionalist behaves like a docile student at the school of Mother Nature, and the liberal – as a child in a toyshop.

4. THE WINDS OF CHANGE: SCIENCE POLICY

We need a special type of science criticism that probes the fundamentals of science in a series of contexts until the limits of their stability are established, looks for incoherences and weak points and builds stocks of possible modifications and generalizations for future use. If epistemological responsibility is to be taken seriously, this inclusive criticism must predict the activating of dormant incoherences and offer strategies to cope with the oncoming dilemmas. Briefly: the inclusive criticism is epistemology in action. It looks like an extremely ambitious enterprise and presents a huge challenge to the scientific community, but after the advent of the quantum theory it is too late to think of retreat.

The inclusive criticism is naturally coupled with a strongly restrictive science policy whose current task is to support the pivotal strategies as opposed to the spread of makeshift theoretical constructs (which can be fabulously successful from the practical viewpoint). Its long-term responsibilities are related to the dilemmas coherence – stability and the eventual bifurcations in science. That sort of restrictiveness can be described as a future-induced censorship.

The inclusive criticism and restrictive policy are expensive programs in the literal sense of the word, but in the long run they are clearly a profitable investment.

The problems concerning the coherence (incoherence) between very large systems of ideas form a complex hierarchy proliferating in all directions (imagine the full scale of the relations mathematics – physics or physics – religion, for example). The intellectual contacts between key aggregates of ideas are thus gradually becoming a full-time job and will probably crystallize into a separate profession, that of the intellectual mediator. Not to be confused with the fashionable interdisciplinary studies, the aim is not mutual enhancement but rather the opposite – reciprocal restraint.

The kind of science policy we advocate here will meet active resistence from the reigning scientific opportunism. The opportunistic attitude is indeed the best possible policy from the individual perspective (and from the perspective of most institutions, as well) but it is definitely not adequate since it ignores all the long-range and long-term effects of what happens here and now.

Finally, the new science policy drops altogether the legend that planning and science do not mix. The policymakers need plans, no matter how risky they may be. Any move in the world of ideas is a hazardous endeavor. We can only dream of a guidebook for the jungle of incoherent ideas. Shall we call it ideonomics or ideodynamics?

5. CONCLUDING REMARKS

Let us make a recapitulation. What should be our reaction to possible splits and bifurcations in science? Support or resistance? The question belongs on the agenda of science policy and no ready-made answers are available. Yet, if a guiding rule is to be articulated, it is this: stick to the conservative scientific values, coherence, stability, cumulativeness; most significantly, they seem sufficient, in their interplay, to make science reinvent itself and admit irreversible bifurcations.

The position of extreme liberalism is an asymptotically conceivable and emotionally appealing vision, incapable of further developing (to meet today's science it can only move backwards). This extremism can be ascribed to the puberty of human reason (clearly detectable in the last pair of centuries) – fierce striving for freedom and, predictably, want of skill to make something useful out of it. Respect for a wider range of values would certainly testify that maturity is going to set in.

In this essay I have tried to inject a bit of coherence into the relationship between traditional rationalism and modern liberalism in the philosophy of science. I hope the text itself is coherent enough and could withstand some inclusive criticism.

Part II

PHILOSOPHY OF PHYSICS

AZARYA POLIKAROV †

THE PROLIFERATION AND SYNTHESIS OF PHYSICAL THEORIES

Tremendous progress has been achieved in the theoretical understanding of entire groups of new phenomena, but the establishment of a unified basis of physics as a whole seems an occupation for the remote future. And this state of affairs has been aggravated by the following development!
Albert Einstein

The step by step conquering of physical knowledge is worth noting. Taken as a whole, such knowledge constitutes the physical world, which includes a number of classes of "phenomena" (and their natures).

For a long period of time physics (and science in general), as is well known, was represented by mechanics, which was regarded as a comprehensive theory of the world, such that reality itself was considered to be mechanical in its essence.[1]

Significant progress in physical (and scientific) knowledge has been made through the transition from one universal (physical, or mechanical) theory to a group of (basic) theories, to which mechanics also belongs (but already with a correspondingly limited domain of validity). This corresponds to the penetration of knowledge into new areas, and the revealing of their particularities.

At this stage the view of the existence of so-called closed physical theories, which are characterized by a particular system of concepts and laws, emerges. Each of these theories gives rise to a ramified system of derivative theories.

Along with this, in every physical domain – mechanics, field physics, etc., alternative theories are framed. Thus we observe a markedly expressed process of differentiation or proliferation of physical theories.

The establishment of inter-theoretical links between these theories, and their syntheses into more general theories, goes hand in hand with this process.

This trend of bridging physical theories as they grow in number (which we will try to elucidate in greater detail) may be considered as a concretization of the more general divergent-convergent course of scientific knowledge in solving scientific problems.

1. THE PROLIFERATION OF PHYSICAL THEORIES.

Closed Theories. In our time the view has been formed that there are several basic independent physical theories that are characterized by specific concepts and laws. Aside from this, they are conceived as complete in principle, i.e., they are not to be

Dimitri Ginev (ed.), Bulgarian Studies in the Philosophy of Science, 53-68.

subject to improvement through minor changes. In this context Heisenherg introduced the concept of "closed (abgeschlossene) theories".[2]

In keeping with this conception Weizsäcker[3] adheres to the following general scheme of (five) mutually connected physical theories, which either exist or are in a state of being framed:

1. The theory of space-time (the special or possibly the general theory of relativity);
2. General mechanics (quantum theory);
3. The theory of the possible types of objects (theory of elementary particles);
4. The theory of irreversibility (statistical thermodynamics);
5. The theory of the universality of physical objects (cosmology).

The theory of elementary particles, which is in a process of formation, has been included along with the existing physical theories. Unlike theories 1, 2 and 4, which are almost completed, the framing of 5 may still be considered an open problem.

Tisza's attempt at systematizing ("structuring") physical theories is of interest.[4] He considers the fundamental physical branches (mechanics, thermodynamics, quantum mechanics) as clusters of deductive systems; or, in other words, the basic discipline is represented by more than one (supplementary) system. In this way he distinguishes among a classical mechanics of particles, a classical mechanics of particles under a microscopic interpretation, and a phenomenological classical mechanics. Analogously, according to him, thermodynamics is represented also by a macroscopic equilibrium thermodynamics, a statistical equilibrium thermodynamics, a macroscopic thermodynamics and a statistical thermodynamics. Along with quantum mechanics, quasi-static quantum mechanics and quantum dynamics are also included here. Some of these systems (quantum dynamics, statistical thermodynamics, and macroscopic thermodynamics) are essentially incomplete.

This system also includes classical gravitational theory, classical electrodynamics, relativistic mechanics and relativistic gravitational theory. Altogether there are eight theories and three other incomplete systems.

Tisza examines the different relations between these systems, namely those of compatibility, controlled inconsistency, and supplementarity, as well as the incompletely understood relation.

Derivative Theories. The basic (closed) theories comprise various branches; derivative theories are built up on their basis. Thus, along with the dynamics of mass points, the system of mass points, and the dynamics of solids, theories have been created covering continuous media (fluids and gases), hydrodynamics, aerodynamics and gas dynamics, the dynamics of deformable solids, elasticity and plasticity, as well as celestial mechanics, perturbation theory, aircraft dynamics, rocket dynamics, hydraulics, the dynamics of machines and mechanisms, and so on.

Field theory as a general trend may be represented by the theory of the electromagnetic field, and the theory of gravitation. From these theories, the electrodynamics of continuous media, cosmology, and so on, are formed.

Within the framework of thermodynamics, the theory of phase transition, the theory of surface phenomena, etc. are created. The thermodynamics of magnetic phenomena, the thermodynamics of superconductivity, the statistical thermodynamics of solids, and so on, are elaborated on the basis of thermodynamics.

Quantum theory embraces: the quantum, many body theory, statistics, radiophysics (electronics), the theory of magnetism, field theory, electrodynamics, the theory of heat absorption, the theory of crystals and solids, of their electricity and heat-conductivity, of their magnetic properties, the theory of semiconductors, of superfluidity, of superconductivity, quantum chemistry, etc.

This process of building up derivative theories is denoted as extensive development of the theory.[5]

Classifications of Physical Theories. The set thus formed of (several dozens of) physical theories is the subject of various classifications. Among the different attempts of this kind we shall note the following approaches (which are of interest in our case):

A) We consider first *Einstein's* approach, where one distinguishes between constructive theories and theories of principle (fundamental theories).[6] To the former belong theories like the kinetic theory of gases, wherein a picture is given of complex phenomena on the basis of relatively simple assumptions. The theories of principle proceed from empirically observed general properties of phenomena, respectively from principles by means of which (mathematical) criteria with a universal validity are formulated. The analytical method is used in them. Examples of such theories are thermodynamics and the theory of relativity.

The constructive theories, according to Einstein, are distinguished by their comprehensiveness, adaptability and clarity; and those of principle, by their logical perfection and the security of their foundations. If any deduction from a theory should prove untenable, the theory must be given up. A modification of it seems impossible without its destruction as a whole.

B) Concerning theories in general (not only physical theories) *Rapoport* distinguishes seven groups,[7] of which we shall consider the first three (which have a bearing on physical theories).

1. Intrinsically mathematical theories, which involve something analogous to the notion of a state, and describe a state-transition mechanism ("equation of motion"), e.g., quantum mechanics.

2. Theories that are like those of the first group except that they contain no state transition mechanism, such as thermodynamics, crystallography, and classical optics.

3. Stochastic theories, which are mathematical and quantized but in which the basis of quantizing is computation rather than measurement; such theories generally do not involve a notion of state, employing rather the notion of an event.

C) In *Madelung's* classification the following (seven) kinds of physical theories are distinguished: point (e.g., point mechanics); quasi-point (mechanics of bodies); continuum (field theory); quasi-continuum (hydrodynamics); combined point and continuum (electronic theory); systemic (thermodynamics); and statistical (in a quasi-point or quasi-continuum form). Aside from these he also speaks of variants - single-point and multi-point theories in a configuration, or in a phase space, respectively.[8]

D) *Miyakishev* provides a detailed table of physical theories according to two basic signs: the kinds of interaction, on the one hand, and the kinds of objects (macroscopic systems and microscopic objects) and levels (macroscopic, microscopic, classical and quantum) on the other.[9]

E) We shall also mention the systematics of physical theories which is based on two signs: the subject under investigation and the approach.[10]

A distinction is drawn between ideal (physicalist) and real (physical) objects. Objects like mass points, absolutely solid bodies and ideal gases are abstract. To the group of concrete objects belong (a) material objects and systems – bodies, molecules, atoms, nuclei, elementary particles, plasma, physical fields; (b) states of matter – aggregate and other states of matter; (c) motion and processes of change of position, form, state, structure, behaviour (function), the basic kinds of interactions, heat, etc.; and (d) space and time.

These objects may be the subject of several basic approaches, such as the dynamic (mechanical), statistical, relativistic, and quantum. Several theories are formed here, for example, mechanics and relativistic mechanics of the mass point, continuum mechanics, quantum mechanics, statistical physics, the theory of gravitation, quantum field theory, and quantum solid-state theory.

According to some contemporary views certain fundamental theories may be regarded as definitively established. A similar attitude toward the final or almost final character of such physical theories is possible within the framework of the (finitist) conception that knowledge of the physical world is reduced to a finite number of types of interaction in nature and a finite number of constants (and their ratios: dimensionless quantities). Correspondingly it will be exhausted through the establishment of a finite number of (fundamental) physical laws or theories respectively.[11]

Along these lines it is considered that the general theory of relativity has such a closed character, provided it has no bearing on elementary particles. This does not rule out the possibility of the development of its mathematical apparatus. [12]

From this point of view it may be assumed that this type of theory is subject to some modification (perfection, generalization) but not to refutation and replacement by another (essentially different) theory. In other words, it is assumed that the modifications in question have no bearing on the theory's character.

The antithesis of this view is expressed by Wigner, who admits the possibility that theories which we consider "true" (relying on a considerably large number of confirming facts) could be as a matter of fact "wrong" because they contradict a more adequate, but not yet discovered, theory.[13]

Alternative theories. Along with the basic theories - as another kind of proliferation - alternative (variant) theories are formed. Examples are Hertz' mechanics, the mechanics of Riemann-MachTreder (R-M-T),[14] and the electrodynamics of Wheeler-Feynman. Aside from the general theory of relativity, at present several alternative theories of gravitation are known, for example, those of Newton-Cartan, Ni, and Belinfante-Swihart. Statistical thermodynamics emerges as an alternative to the usual (phenomenological) thermodynamics. As regards quantum mechanics (irrespective of the variants with hidden parameters), the theory not only appears in two variants – as matrix and wave mechanics – but these variants retain a certain relative independence.[15]

The idea of a plurality of physical theories, in particular of the possibility of alternative theories, came into being in the 19th and 20th centuries. Thus in the last century Herschel wrote "nothing is more common in physics than to find two, or even many, theories maintained as to the origin of a natural phenomenon."[16]

Of particular importance in this respect is Hertz's mechanics. It has contributed to a change in the general epistemological attitude of leading physicists. Hertz himself recommends a thorough consideration of the existing possibilities in all directions. Moreover, he is fully aware that "one image may be more suitable for one purpose, another for another; only by gradually testing many images can we finally succeed in obtaining the one that is the most appropriate."[17]

A marked turn in this respect is connected with the construction of variant electrodynamics by Wheeler and Feynman. On the basis of this accomplishment Feynman proclaimed the plurality of the possible schemes for interpreting nature as one of the amazing peculiarities of its cognition. He dwells on this question in his Nobel lecture.[18] Pointing out that there exists no scientific method for determining which of two theories based on profoundly differing ideas but equivalent from a mathematical point of view is the true one, Feynman emphasizes that they are nevertheless not identical as regards their possibilities of modification and further development. Such a modification may be quite natural for one of these theories and unnatural for the other. He therefore thinks that a good theoretical physicist today might find it useful to have a wide range of physical viewpoints and mathematical expressions of the same theory available to him.

This view is propounded also by a number of philosophers of science.[19] For instance, according to Feyerabend the plurality of theories must not be regarded as a preliminary stage of knowledge, which will at some time in the future be replaced by the One True Theory. He assumes that theoretical pluralism is an essential feature of all knowledge that claims to be objective. The road to experimental test passes through the elaboration of competing views.[20]

The elucidation of the relationships between alternative theories presupposes an appropriate analysis of their basic concepts, principles and mathematical apparatus. Thus, for comparing the theories of gravitation, Thorne et al. framed the general foundations (the theory), consisting of a glossary of fundamental concepts, the delineation of the different theories, the establishment of the validity of the principle of equivalence, and so on.[21] A theory, according to these authors, can be regarded as the equivalence class of all its (mathematical) representations.

The processes of proliferation of physical theories can be summed up and presented graphically through the following table:

Basic theories	*Derivative theories*
Mechanics	Hydrodynamics Theory of Elasticity ...
Field Theory (Electrodynamics)	Electrodynamics of Continua ...
Thermodynamics	Theory of Phase Transitions ...
Quantum Theory	Quantum theory of magnetism Quantum theory of superconductivity
Alternative Theories	Derivative alternative theories
Hertz' mechanics R-M-T mechanics Feynman's electrodynamics Newton-Cartan's theory of gravitation Ni's theory of gravitation Belinfante-Swihart's theory	

It is clear from this table that considerable extensive development of the basic theories is present, whereas with respect to the alternative theories such an elaboration of derivative theories has not taken place. In our opinion this is not simply a blank spot on the map of theoretical physics, but we think it may prove to be a promising area for the creative activity of theoretical thought (and possibly for the obtaining of non-trivial practical results).

2. THE SYNTHESES OF PHYSICAL THEORIES

The unifying trends in theoretical knowledge are manifested with respect to the concepts, hypotheses (premises), problems, methods, mathematical apparatus, laws, principles and the theories themselves, a point on which we shall dwell in what follows.

The unity of the physical world is supported by facts such as those concerning the existence of the so-called large numbers (dimensionless ratios), e.g., 10^{40}, by which the size of the Universe (Metagalaxy) is expressed in elementary lengths ($= 10^{-13}$ cm), and also by those concerning the time of existence of the Universe in units of elementary time ($= 10^{-23}$ sec), and the relation of the strong and gravitational interaction between elementary particles, or the ratio of the mass of the universe to the mass of the proton, which is 10^{80} ($= (10^{40})^2$).[22]

The similarity in general appearance of different excited systems – condensed matter, atoms, atomic nuclei and elementary particles – which is manifested in spectra of the same type in a wide energy range, is striking.[23]

The Link Between the Physical Concepts. Along with the specific (basic and other) concepts one also encounters concepts that are common to particular theories and that are sometimes denoted as "borrowed". As a rule every succeeding physical theory incorporates concepts from previous theories. For instance a number of theories operate with such kinetic concepts as length, time, velocity, and dynamic concepts such as mass, energy, action, etc.

It is worth recalling the transition from the concepts (quantities) of classical mechanics to those of quantum mechanics by an appropriate generalization. According to Bohr – and this is an essential aspect of his complementarity conception – the classical (macroscopic) concepts are in a certain sense universal.

The conception of complementarity admits that: non-classical theoretical notions underlie an empirical interpretation in classical terms, however different from the classical objects they may be and notwithstanding the fact that the classical theory is not applicable to them; it is a universal statement that their connection with the facts is expressed in the language of classical physics (mechanics and electrodynamics). This is due to the circumstance that the measuring instruments are classical. To put it in another way, the language of classical physics is also valid for the non-classical domain (i.e., it is a universal language).

Theories are associated with some general hypotheses of the same sort as the axioms defining the character of Euclidean space, which does not change with time (i.e., the scales are invariant). Concerning a particular group of (non-relativistic) theories it may be assumed that $c=\infty$ and for another (non-quantum) group, that $h=0$.

One can also speak about the general character of the problems of physical theories, especially (after I. Akchurin) of the following problems: the selection of the abstract space of a theory (Zenonian-Eudoxian problems); the determination of universal operational procedures (Descartes-Einsteinian problems); and the search for an elementary object (eleatic-Heraclitean problems) and universal laws of motion (Galilean-Newtonian problems).[24]

Certain methods (e.g., those of mechanics) are also distinguished by their generality; they find application in many other areas (optics, electrodynamics, theory of relativity, statistical physics, and quantum theory).

A similar thing can also be said with respect to laws, e.g., the energy conservation law holds good for all physical theories.

The unity and general character of the laws finds its expression in their symmetry and invariance. The idea of symmetry is a unifying concept that underlies the invariance of laws under a group of transformations. The principle of invariance – as has been noted by Wigner – provides a set of correlations between the laws of nature in the same way as the physical laws provide the correlation between events.[25]

The concept of symmetry has come to be a leading idea in physics since the theory of relativity was accepted. L. Radicati di Brozolo has devoted a paper to this topic, in which he justifiably emphasizes that the requirement of invariance as a way to determine how the explicit form of an interaction becomes an integral part of the "philosophy" of physics.[26] Along these lines the same author maintains that the search for symmetry becomes one of the most powerful tools for the discovery of new laws, and one of the surest guiding principles in the endless struggle to achieve a logical and harmonious description of nature.

Graphically, the relationship between symmetry and invariance on the one hand, and the law of energy conservation and other consequences on the other, may be represented by the following scheme:[27]

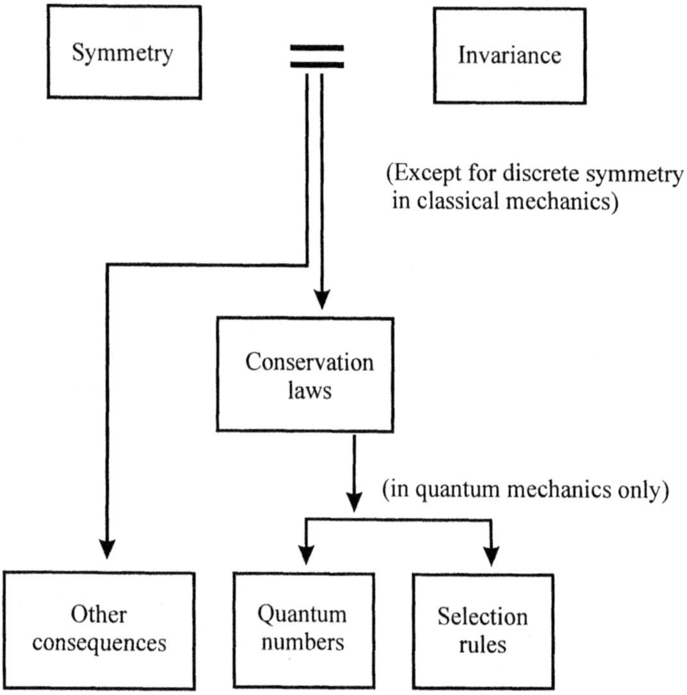

This idea is supplemented by the idea of a spontaneous breakdown of symmetries (in the theory of elasticity, the dynamics of fluids and superfluids, etc.).

In other cases one can speak about the preservation of the form of the laws. One has in mind the fact that in different theories – such as classical mechanics and special relativity – certain laws, e.g., Newton's second law, are written in the same general form.

Let us also note certain conceptions of a general character, such as the atomistics that permeates a number of theories.

General Physical Principles – The Correspondence Principle. With respect to principles we shall note the ample validity of the variational principles, of the principle of relativity, etc. for some class of theories. For example, the variational principles are applied in classical dynamics, as well as in field theory and quantum mechanics.

The metanomological statements also belong here; these statements include:

1. *Factual* metanomological statements like "Newton's laws of motion are invariant under time reversal".

2. *Normative* metanomological statements, such as "The equations of motion are not to depend on any frame of reference".

3. *Methodological* metanomological statements like "Law statements should not include egocentric particulars such as 'now'".[28]

For a sequence of physical theories the correspondence principle has been formulated. According to Bohr this principle represents a supplement to his theory (of atomic spectra), permitting it to provide a more complete explanation of the characteristics of emission. This occurs by establishing the way in which the classical theory (Maxwell-Lorentz's electrodynamics) may be used.

To the credit side of this principle, it may be noted that it has served as an "Ariadna's thread" in the formulation of matrix mechanics. In this way the validity of the principle was also extended to the relationship between Bohr's (old) and the new quantum mechanics, as well as between the latter and classical mechanics.

A statement of Hund is interesting in this respect. He has said that "the term 'correspondence principle' was later used in two senses: in a narrower sense as a determinant of intensities and as a guiding principle (*Richtschnur*) for the establishment of selection rules, and in a broader sense as a general indicator of how to establish the quantum-theoretical relations in such a way as to pass asymptotically to the classical ones."[29]

At the same time it is to be noted that in Schrödinger's variant of quantum mechanics the necessity of this principle falls away. Schrödinger emphasizes that wave mechanics makes any kind of principles of correspondence and selection superfluous.[30]

Independent of this, there is a generalized version of the correspondence principle encompassing the relationship between the theory of relativity and classical mechanics, between the special and the general theories of relativity, and in general between successive theories in a given domain or of a definite subject matter.[31]

The application of the principle and the establishment of a correspondence relation between successive theories encounters difficulties. In various cases, as Bunge indicates, the non-relativistic limits sometimes do not exist and at other times are multiple; the asymptotic approximation of a new theory may not coincide with the old one; the classical limits of quantum theory are not well known, etc.[32]

In the literature a more general (philosophical) conception of the continuity of successive theories, or of the inclusion of a preceding theory as a limiting case in its successor, has also been maintained.[33] Here one bears in mind the generalization of theories not only with respect to the quantitative results, but also with respect to their qualitative contents. Such a conception cannot have but a restricted validity.

Taking into account the different conceptual contents of superseding theories, the thesis of their incommensurability has been proclaimed. Without discussing this problem in detail we may say that this thesis also has an altogether restricted validity. In general, the relations between different successive theories occupy the range from correspondence to incommensurability. Hence, in the general case we may have to deal with a quantitative correspondence, and in the particular, with a qualitative one.

The Synthesis of Theories. One can – and does – speak about the synthesis of theories in a different sense, which can be given the following rank-ordering:

– the generalization of a given theory (and the inclusion of its predecessor in it);

– the establishment of the unity of two theories;

– the obtaining of a theory by combining two theories, or

– the merging of two theories into a more general (unified) one, and, finally, into one (superunified) theory;

– the inclusion in a synthesis of this kind of scientific theories beyond the domain of physics.

Laue draws attention to the recurring situations in physics where two hitherto completely independent physical ideas created by different groups of people, e.g., optics and thermodynamics, or wave theory and X-rays, suddenly meet and spontaneously merge.[34]

This is obvious from the table that Weisskopf provides with regard to the unifying principles of physics, as reproduced below:[35]

Principle	*Author or theory*
Unity of natural laws in heaven and earth	Newton
Unity of heat and mechanics	Mayer, Helmholtz, Joule
Unity of electricity, magnetism and optics	Faraday, Maxwell
Unity of space, time, matter and gravity	Einstein
Unity of physics, chemistry and material science	Quantum mechanics
Unity of atomic, nuclear and subnuclear phenomena (?)	Field theory (?)

A typical synthesis in physics was achieved by the theory of relativity, on the basis of which mechanics and electrodynamics were unified. An analogous synthesis between physics and geometry was achieved in general relativity.

The relativistic quantum mechanics of Dirac, in which a partial synthesis of quantum mechanics and the (special) theory of relativity was attained, also deserves mentioning.

It is well known that Einstein pursued a programme aimed at the creation of a unified field theory, which was to include the theory of electromagnetism and the theory of gravity. On this basis he hoped to achieve a spatial-temporal explanation of the nature of the electrical charge.

The most considerable success in this direction has been attained in recent years by the framing of the grand unified theory of the electromagnetic, weak and strong interactions (cf. e.g. J. Goldstone, P. Higgs, S. Glashow, S. Weinberg, A. Salam, G. t'Hooft). It has passed through several stages: the so-called eightfold way was elaborated and the hypothesis suggesting the existence of quarks was advanced, both in the 1960's, and the conception of the existence of neutral currents was confirmed in the 1970's.[36] A visual representation of the lines of inquiry into strong and weak interactions is provided in the scheme reproduced below.[37]

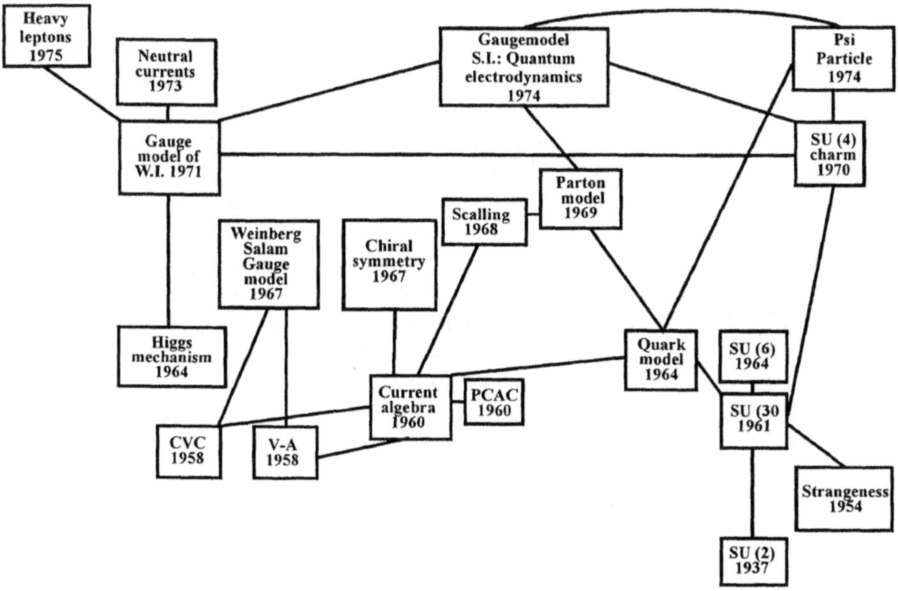

In our time, due to successes in the synthesis of fluctuation theory and general relativity, the problem of black holes emission was elucidated, and a new type of particle generation was discovered (S. Hawking, 1974).[38] This allows one to speak of a reinforcing of the unitarian trend in recent physics.

In nuclear physics the so-called super-symmetry was introduced, which included bosons and fermions. There is analogous work on the creation of a quantum theory of gravitation, or of a supergravitation theory, from which it follows that along with gravitons (bosons) there should also exist gravitino particles (fermions).

Nevertheless, the problem of the synthesis of the two basic trends of contemporary theoretical physics – quantum theory and the theory of relativity – and the building of a quantum theory of gravitation in particular, remains open.

The distinction made by Branskij of the non-fundamental theories into complex and hybrid theories is worth noting.[39] When a non-fundamental law is deduced from two fundamental theories we have to deal with a complex (non-fundamental) theory. Magnetohydrodynamics may serve as an example of such a theory. We deal with a hybrid theory in the case where the application of two theories leads to a partial explanation of the existing empirical data and a partial prediction of new data, and where this application also leads to paradoxes. Some mechanical theories for explaining electromagnetism have had such a character.

The synthesis of physical theories is manifested in the general mathematical apparatus of various theories. Already in Maxwell's time, and thanks to his work, it became clear that in various branches of science we may deal with systems of magnitudes in which the mathematical form of connection is the same irrespective of the fact that their physical nature may be totally different.[40] Along these lines Boltzmann points out that the same differential equations are valid for strikingly different phenomena. Thus heat conductivity, diffusion, and the propagation of electricity in conductors occurs according to the same laws. The same equation may be conceived as a solution to problems in hydrodynamics and in the theory of the potential. The fluid vortex theory, like the theory of friction in gases, shows a surprising similarity to those of electromagnetism, etc.[41] From this point of view it is worth noting that the different kinds of interaction may be represented by Lagrangians.

In close connection with the establishment of symmetries of physical laws, group theory acquires an essential and ever increasing relevance. If in geometry this is associated with Klein's famous Erlangen programme, in physics it has its outset in Einstein's theory of relativity. Group theory – as di Brozolo puts it – became the best guide for discovering the form of the dynamical laws.

In the contemporary unified theory of elementary particle interactions, different groups: SU(2) x U (1), SU(3), were applied, and their union was achieved. The representation via the SU(5) group is considered the most adequate.[42] One can get an idea of this union by considering the following scheme.[43]

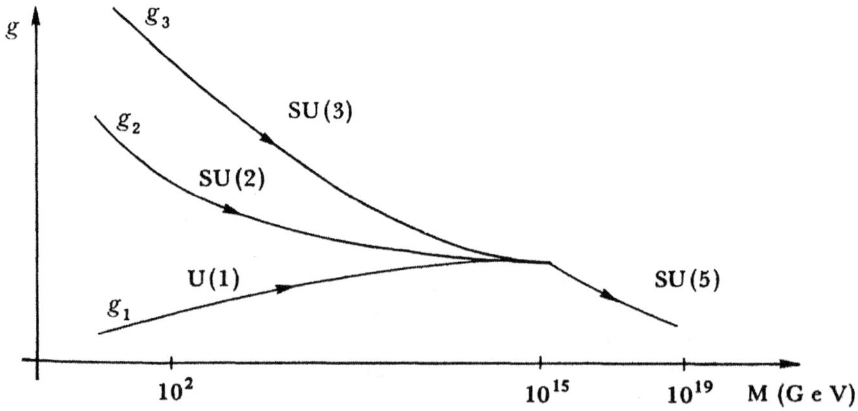

The progress of physics is associated with the extension of those groups of transformations on which the physical theories are based.

It has recently been established that the same mathematical formalism is applied in the theory of phase transitions (of the second order) as in quantum chromodynamics. This leads to the idea of an asymptotic freedom of the quarks, i.e. to their being free only over very small distances.

The idea of synthesizing physical theories in a unified or superunified theory is supported by various authors. As early as in 1908, Planck arrived at the conclusion that a unification (*Vereinheitlichung*) of theoretical physics is characteristic of its entire development.[44]

According to a statement of Einstein,

> From the very beginning there has always been present the attempt to find a unifying theoretical basis for all these single sciences, consisting of a minimum of concepts and fundamental relationships, from which all the concepts and relationships of the single disciplines might be derived by a logical process. That is what we mean by the search for a foundation of the whole of physics. The confident belief that this ultimate goal might be reached is the chief source of the passionate devotion which has always animated the researcher.[45]

With a view to achieving a unity of the closed theories adduced earlier according to Weizsäcker with the postulates on which quantum mechanics is based (9 in number) he advances five further postulates, viz., those of the approximately static cosmology, of the existence of ultimate objects, interaction, symmetry and expansion.[46]

The question of the possibility of attaining a synthesis between theoretical physics and biology is also discussed in the literature.[47] In connection with this, some considerations of a general nature are appropriate. It might be expected that a synthesis is in principle possible given certain conditions concerning the homogeneity of the theories. Such a condition is, for example, that the theories belong to the same class in a given classification, e.g., in that of Rapoport (the relevant theories all having or not having an equation of motion), or in that of Miakishev (the theories all being either macro- or microscopic), and so on.[48]

Another consideration of this kind is the following. Bearing in mind that the principles or postulates of the examined physical theories may be physical statements, or sometimes metastatements or metanomological statements, in juxtaposing two theories it is necessary to take into account the nature of these postulates. Statements should bejuxtaposed with statements, and metastatements with metastatements without mixing the two types of principles.[49]

Finally, the unification of theories, in accordance with our classification given earlier, could be effected by a generalized approach, or by a combination of an object and generalized approach.

There are competing programmes, or trends, whose goal is to attain a unification of physical theories. The possibility of alternative unifying theories, i.e., of the bringing about of various kinds of synthesis, may be admitted.[50]

As a consequence of the above considerations, one can speak about three types of proliferation of physical theories: basic, derivative and alternative. There are various forms of connection among them (between concepts, laws, principles, etc.), and various forms of synthesis into more general theories.

This dialectical combination of opposite processes may be regarded as a concretization of the divergent-convergent approach to solving scientific problems in general.[51]

The following problems remain open: a) the fourth form of proliferation of physical theories (derived from the alternative ones); b) the framing of alternative unified theories; and possibly c) the framing of super-unified physical theories; and finally d) the synthesis that goes beyond the framework of physics.

NOTES

[1] It would be more correct to speak of a ramified system of mechanical theories, which is obtained through the application of mechanics to particular domain or subjects.
[2] Heisenberg, 1971: 87f.
[3] v. Weizsäcker, 1971: 233-234.
[4] Tisza, 1965: 62.
[5] Kuznetsov and Pechenkin, 1973: 551.
[6] Einstein, 1973: 228; Einstein, 1950: 54.
[7] Rapaport, 1958: 927, after Suppe, 1974: 123.
[8] Madelung, E. 1957: 1.
[9] Myakishev, 1973: 111.
[10] Polikarov [b], 1979: 419.
[11] Feynman, 1965: 172.
[12] Kompane'ets, 1967: 11, 46.
[13] Wigner [b], 1970. As far as general relativity is concerned, Weinberg admits that there may be a whole world of new physics, and that general relativity does not provide an adequate framework for understanding the physics of these superhigh energy degrees of freedom [Cf. Weinberg, (1980): 1216.]
[14] Treder, 1972.
[15] Dirac, 1967: lect. 1.
[16] Herschel, 1968: 100.
[17] Hertz, 1956.
[18] Feynman, 1968: 44.
[19] Cf. Polikarov [a], 1979.
[20] Feyerabend, 1965.
[21] Cf. Thorne, Lee and Lightman, 1973: 3563.
[22] Cf. Dirac, 1973: 45.
[23] Cf. Bromley, 1980: 110. A. Mercier notes the regular behaviour of different types of clocks as one of the most remarkable facts of the physical world. This speaks in favour of the universality of time as a magnitude upon which other physical quantities depend. Cf. Mercier, 1971: 285.
[24] Cf. Akchurin, 1973.
[25] Wigner [b], 1970
[26] Radicati di Brazolo, 1979: 524.
[27] Yang, 1964: 394.
[28] Bunge 1961: 518.
[29] Hund, 1967: 75.
[30] Schrödinger 1926: 1049.
[31] While with respect to the transition from classical to quantum theory this principle is a heuristic one, in the more general case of establishing the relation of correspondence between superseding theories it no longer has this character.
[32] Bunge, 1973: Ch. 9.
[33] Krajewski, 1977. Along these lines one believes that the correspondence principle is in its essence an epistemological principle (cf. Pretrov and Omarova, 1981).

[34] v. Laue, 1950: 14-15.
[35] Weisskopf, 1979: 241.
[36] Salam, 1980: 723.
[37] Crane, 1980: 47.
[38] Sciama, 1979: 717.
[39] Branskij, 1973: 38. He too speaks of fragmentary theories. They are built upon a non-fundamental law deduced from a fundamental theory. Such a fragmentary theory is, e.g., the theory of Brownian movement of Einstein and v. Smoluchowsky.
[40] Maxwell, 1890.
[41] Boltzmann, 1919: 7.
[42] Georgi, 1980: 649.
[43] Iliopoulos, 1979: 113.
[44] Planck, 1943: 4.
[45] Einstein, 1973: 324.
[46] v. Weizäcker, 1971: 264-275.
[47] Cf. Bohr, 1958: 5; and Wigner [a], 1970: 35.
[48] Proceeding from Einstein's classification, J. Bub and W. Demopoulos speak about a third, mixed type of theory (both constructive and of principle), classical statistical mechanics being an instance according to them. Cf. Bub and Demopoulos, 1973: 92.
[49] Polikarov, 1965: 31.
[50] For instance, the grand unified theory in point of fact represents a large class of such theories (cf. Weinberg [b], 1970: 1217). Attention is to be paid to the opinion of Glashow, that the standard theory will not survive as a correct and complete picture of physics: Glashow, 1980: 1319.
[51] Cf. Polikarov, 1966, and Polikarov, 1977; as well as Popper, 1972.

REFERENCES

Akchurin, A. I. "Some Regularities in the Development of Knowledge and the Problem of its Synthesis", *Synthesis of Contemporary Scientific Knowledge* (in Russian), Moscow, 1973.

Branskij. V. P. *Philosophical Foundations of the Problem of the Synthesis of Relativistic and Quantum Principles* (in Russian), Leningrad, 1973.

Bohr, N. *Atomic Physics and Human Knowledge*, New York, 1958.

Bromley, D. A. "Physics." *Science* 209 (1980).

Bub, J. and Demopoulos, W."The Interpretation of Quantum Mechanics." *Logical and Epistemological Studies in Contemporary Physics*, R. Cohen and M. Wartofsky (eds.), Dordrecht, 1973.

Bunge, M. "Laws of Physical Laws." *Am. J. Phys.* 29 (1961).

Bunge, M. *Philosophy of Physics*, Dordrecht, 1973.

Crane, D. "An Exploratory Study of Kuhnian Paradigms in Theoretical High Energy Physics." *Soc. Stud. Sci.* 10 (1980).

Dirac, P. A. M. *Lectures on Quantum Field Theory*, New York, 1967, lect. 1.

Dirac, P. A. M. "Fundamental Constants and their Development in Time." *The Physicist's Conception of Nature*. J. Mehra (ed.), Dordrecht, 1973.

Einstein, A. *Out of My Later Years*, New York, 1950.

Einstein, A. *Ideas and Opinions*, London, 1973.

Feyerabend, P. "Problems of Empiricism." *Beyond the Edge of Certainty*. R. Colodny (ed.), Englewood Cliffs, N.J., 1965.

Feynman, R. *The Character of Physical Laws*, London, 1965.

Feynman, R. "The Development of the Space-Time View of Quantum Electrodynamics." *Physics Today* 19 (1968).

Georgi, H. "Why Unify?" *Nature* 288 (1980).

Glashow, S. L. "Toward a Unified Theory..." *Science* 210 (1980).

Heisenberg, W. *Schritte über Grenzen*, München, 1971.

Herschel, J. F. W. "Scientific Method," *Science: Men, Methods, Goals*, B. Brody and N. Capaldi (eds.), New York, 1968.

Hertz, H. *The Principles of Mechanics*, New York, 1956.

Hund, Fr. *Geschichte der Quantentheorie*, Mannheim, 1967.

Iliopoulos, J. "Toward a Unified Theory of Elementary Particle Interactions." *Einstein Symposion Berlin,* H. Nelkowski et al. (eds.), 1979.
Kompane'ets, A. S. *Could There Be an End to Physics?* (in Russian), Moscow, 1967.
Krajewski, W. *Correspondence Principle and Growth of Science,* Dordrecht, 1977.
Kuznetsov, V. I. and Pechenkin, A.A. "The Formation of Quantum Chemistry." *Synthesis of Contemporary Scientific Knowledge* (in Russian), Moscow, 1973.
Laue, M. *Geschichte der Physik,* Bonn, 1950.
Madelung, E. *Die mathematischen Hilfsmittel des Physikers,* Berlin, 1957, Tl. II.
Maxwell, J. C. *Scientific Papers,* Cambridge, 1890.
Mercier, A. "Thoughts on the Dynamics of Foundations, or What I Believe." *Foundations of Physics* 1 (1971).
Myakishev, G. Ya. *Dynamical and Statistical Laws in Physics* (in Russian), Moscow, 1973.
Planck, M. *Wege zur physikalischen Erkenntnis,* Bd. I, Leipzig, 1943.
Polikarov, A. "Logical Solution Structures of Scientific Problems." (in Russian). *NDVSh. Filosofskie nauki* 1 (1965).
Polikarov, A. *Relativity and Quanta* (in Russian), Moscow, 1966.
Polikarov, A. *Problems of Scientific Knowledge* (in Bulgarian), Sofia, 1977.
Polikarov, A. [a] "Concerning the Plurality of Scientific Theories," (in Bulgarian). *Filosofska Misl* (1979).
Polikarov, A. [b] "Concerning the Classification of Physical Theories." (in Russian). *C.R. Acad. Bulg. Sci* 32 (1979).
Popper, K. *Objective Knowledge,* Oxford. 1972.
Pretrov, Yu. and Omarova, S. A. "On the Epistemological Nature of the Correspondence Principle." (in Russian). *Vestnik Moskovskogo Universiteta,* ser. 7 Filosofia, 1981.
Radicati di Brazolo, L. "Einstein and the Role of Symmetry in Modern Physics." *Relativity, Quanta, and Cosmology in the Development of the Scientific Thought of Albert Einstein.* Fr. de Finis (ed.), New York, 1979, Vol. II.
Rapaport, A. "Various Meanings of Theory." *Am. Political Science Rev.* 52 (1958).
Salam, A. "Gauge Unification of Foundamental Forces." *Science* 210 (1980).
Schrödinger, E. "An Undulatory Theory of the Mechanics of Atoms and Molecules." *Phys. Rev.* 28 (1926).
Sciama, D. W. "Black Holes and Fluctuations of Quantum Particles." *Relativity, Quanta, and Cosmology in the Development of the Scientific Thought of Albert Einstein.* Fr. de Finis (ed.), New York, 1979, Vol. II.
Suppe, F. "The Search for Philosophic Understanding of Scientific Theories." *The Structure of Scientific Theory,* F. Suppe (ed.), Urbana (Ill.), 1974.
Thorne, K. S., Lee, D. L. and Lightman, A. P. "Foundations for a Theory of Gravitation Theories." *Phys. Rev.* D 7 (1973), 12.
Tisza, L. "The Logical Structure of Physics." *Boston Studies in the Philosophy of Science.* M. Wartofsky (ed.), Dordrecht, 1965.
Treder, H.-J. *Die Relativität der Trägheit,* Berlin, 1972.
Weinberg, S."Conceptual Foundations of the Unified Theory of Weak and Electromagnetic Interactions." *Science* 210 (1980).
Weisskopf, V. F. "Contemporary Frontiers in Physics." *Science* 203 (1979).
Weizsäcker, C. F. *Die Einheit der Natur,* München, 1971.
Wigner, E. [a] "Physics and the Explanation of Life." *Found. Phys.* 1 (1970).
Wigner, E. [b] *Symmetries and Reflections,* Bloomington-London, 1970.
Yang, C. N. "The Law of Parity Conservation and Other Symmetry Laws of Physics." *Nobel Lectures in Physics 1942-1962,* Amsterdam, 1964.

MICHAEL BUSHEV

ON HUMAN AGENCY IN PHYSICS

> *The drama of contemporary mind is that, having solved the mystery of the Universe, it replaced it with another one - the mystery of itself.*
> *Alexandre Koyré*

1.

In what way does the problem of mind appear in physics? What made Roger Penrose state:

> Mind is part of our Universe and any physical theory, which does not duly incorporate it, drops hopelessly behind the adequate description of the world .

(Penrose, *Shadows*). Is mind (thinking, understanding, rationalization, awareness, insight, self-reflection, introspection, etc.) peculiar only to animate nature? And physics itself is it not a science of inanimate nature?

2.

Mankind did not always divide nature into animate and inanimate. Primeval people believed that all objects, plants and animals are endowed with a soul. Animism (from *anima* - soul, spirit) is the belief that natural phenomena can and should be explained in the same way and by the same "laws" that govern the subjective - deliberate and purposeful - human activity. Animism attributes conscious life to nature and natural objects. Animistic (and anthropomorphic) concepts point at living beings, man and mankind as the ultimate goal of universal evolution.

Later animism turned into pantheism (considering Nature as God), which was superseded by the politheistic belief and, finally, monotheism (and atheism) dominated.

Animism and religion gave man the confidence that in Nature there is a supreme goal, a cosmic motive power, a mastermind - the *nous* (of Plato and Aristotle). Aristotle was interested not in the question *how? (in what way?),* but in the question *why? (for what purpose?*). All Nature, according to Aristotle, is built upon the existence of *causes finales* (the ultimate causes, the purposes for which something had been created), they gave the key to our knowledge of Nature. Processes in Nature resemble biological processes, they are governed by a common strict design.

Dimitri Ginev (ed.), Bulgarian Studies in the Philosophy of Science, 69-82.

In the 17th century, when Kepler, Galileo and Newton set the foundations of modern natural science, Nature was still looked upon as a divine creation. Kepler concluded his "*Cosmic Harmony*" with the excited words:

> Thank you Lord, our Creator, for allowing me to set my eyes on the beauty of Your creation and to rejoice at the grandeur of Your deeds.

3.

Modern science germinated on the conflict of two types of rationality - Aristotelian and Galilean. Galileo won, Aristotle lost. Science rejected the vitalistic views and the Aristotelian *goal-oriented cause.* To Galileo there was no room for the question *why?* in science. Confusion of religion with science was inadmissible. When Napoleon asked Laplace what was the position of God in his system of the world, the scientist answered with dignity: " Your Majesty, I don't need such a hypothesis".

From Galileo's time man excluded himself from the nature studied by him. For the method of objective description the observer remained beyond the frame of this description. Man, as a matter of fact, has striven to take the standpoint of the Creator of nature. To dominate over it.

Dwelling on the objective method, Jacques Monod, one of the founders of molecular biology and a 1965 Nobel-prize winner for medicine and physiology, wrote in his book "*Chance and Necessity*":

> The corner-stone of scientific method is the postulate of objectivity. . . the systematic refusal of any interpretation of phenomena by means of *causes finales*, that is through a "design". There can be no genuine knowledge without the postulate of objectivity. (Monod, *Chance*, 83).

The discovery of the principle of objectivity has an exact birthday - the formulation of the principle of inertia (Galileo) by which the foundations were laid not only specifically of mechanics but also of scientific method in general.

However, along the lines of objectivity, man deliberately excluded himself and his human values from the picture of the physical world. The result was rather discomforting.

> The ancient alliance, wrote J.Monod (*opus cit.)*, exists no more. Man had to wake up from his millennial sleep in order to realize his absolute loneliness, his fundamental detachment from nature. . . At last he found out that he is alone in the boundless open spaces of the indifferent Universe in which he came into being at the whim of chance.

Bertrand Russell (Russel, *Science*) was equally explicit by stating:

> Man is the product of causes which had no idea of the final result. . . Science does not treat the question of values, they have no relation to truth or untruth.

"Overwhelmingly hostile" Steven Weinberg (Weinberg, *The first*) called the Universe and continued:

> The more the Universe seems comprehensible, the more it also seems pointless.

4.

The painful dilemma - between the rationality of objective science and the irrationality of the "*ancient alliance*" of man with Nature - became particularly acute in the 19th century when the conflict burst forth between the biologists sharing the evolutionist view and the Church. The centre of the conflict was the *argument from design,* which in the final analysis boiled down to the old dispute over the existence of God. According to the argument from design if there is a watch there should also be a watchmaker.

To Kepler's thinking, in astronomy the argument from design was only too natural. Until the 19th century both Church and scientists agreed that it held for biology too. Darwin's theory of evolution revealed that in the animate world processes of natural selection run, thanks to which, chance variations of hereditary features determine the particularities of biological species throughout long periods of time.

Modern science proved beyond any doubt the correctness of the evolutionist doctrine. The discovery of the structure and the functioning of DNA demonstrated that the mechanisms of evolution are consequences of physical and chemical laws. It is these laws that do the work of the designer. For biology the argument from design is dead. The great achievements of (micro-)biology became possible owing to the reduction of the complex behavior of living beings to the simplest behavior of the building molecules. The reduction of complex to simple, of the apparently purposive behavior of organisms to the purely mechanical properties of the constituents convinced microbiologists that living cell is in its way a "chemical machine".

Along with the "mechanization" of the views on the structure and functions of living organisms, biology abolished the idea of purposefulness of their evolution and, of course, of value as the goal of evolution. Purpose and value were expelled from the objective description of living nature, they were chased away from scientific knowledge.

> "Any confusion of knowledge with values is unlawful, forbidden."- wrote J. Monod (*opus cit.*).

The thesis of a pointless in general world is equally well accepted in sciences of inanimate as well as of animate nature.

5.

From the point of view of physics the mechanical models that help molecular biology to find out the structure and the functions of nucleic acids and of enzymes belong rather to the 19th century. It was in the 19th century that physicists tried to explain everything (the electromagnetic field included) with the help of mechanical models, and for all uncomprehended phenomena readily introduced certain mechanical carriers (ether, caloric, phlogiston, etc.). Biology followed the same path and introduced at one moment the hypothetical *vis vitalis* (the vital force), at another the hypothetical *gemmule* (the heredity substance), and so on.

Classical science left its enduring imprint on the style of thinking in the sciences of the animate world: from biology to humanities and to a great extent in fine arts. It is a curious fact that while for physics itself this style of thinking stopped at the threshold of 20th century, in extraphysical views on the world it preserved a number of its typical features.

The classical picture of the world was started by Galileo with his appeal: to isolate the processes from their natural bonds, to then describe them mathematically and in such a way to explain them. To Newton the world was not a God's creation any more that was to be grasped only in its wholeness. Science began to see the world as independent not only from God, but also from man. The ideal emerged of the objective description and explanation of nature. The distinct aspects of natural processes were submitted to a due isolation, thus an objective observability and understanding being achieved of the laws governing them. Next, the mathematical relations were sought between these processes and through them laws were obtained of which it was accepted that they retain universal validity throughout the whole cosmos.

The method of classical natural science determined the grandiose successes of mechanics in 18th century, of optics and thermodynamics in the first half of the 19th century. The power of the method expanded its range of application, while its typical features achieved clear-cut and precise outlines.

The *ideal* of the classical method is simplicity, transition to the simplest possible physical objects: elementary bodies, elementary motions, elementary forces, in other words objects of the smallest possible number of characteristic features. *Simplex sigillum veri* ("simplicity is the hallmark of truth") is the motto expressing this ideal.

The *core* of the method is the reduction (the analysis, the partition, the decomposition) of processes and phenomena, followed by inductive description and classification.

The *principle* of the method is objectivity. The scientist is an unprejudiced and impassive observer, a recorder of events who remains "out of the brackets" of the cognition process. (It is small wander that the idea of creativity in science is at odds with the principle of objectivity.)

The classical method precludes the human (anthropomorphic) agency as well as axiological reasoning and judgments. Along with the abolition of the "ancient alliance" the classical method doomed man to a complete alienation from Nature. While Nature itself was seen as an automaton about which someone (logical thinking? God? Laplace's demon?) might know in principle the exact causes and the complete consequences, and they would be forever imprinted in Plato's world of ideas, where there is no room for human imperfection. In this picture the scientist himself operates as an immaterial observer of Nature who is as much a stranger to it as the watchmaker is to the watch.

> "Thus the bold and attractive hypothesis", wrote I.Prigogine and I.Stengers (Prigogine et Stengers, *La Nouvelle*), "turned into a "sad" reality. Every new success of natural science increases the uneasiness and the feeling of estrangement. Nature is featureless

and indifferent, while man is reduced solely to the cognizing consciousness which in itself is impenetrable to mind."

6.

At the beginning of 20th century. the theory of relativity and the quantum theory appeared in physics almost simultaneously and these two theories drastically changed classical views.

Newton's laws do not presume the existence of an observer with whatever physical properties. In these laws objectivity is identified with the absence of an observer from the description of the motions of bodies. The theory of relativity took the observer out of his anonymity by highlighting his special part in the definition of simultaneity. This, naturally, does not imply "subjectivist" physics but a physics submitted to those limitations that make man part of the physical world.

The ideas of relativity changed the view of objectivity in classical physics but left intact its model of a "complete" description of Nature. Quantum theory shattered this model too. It was found that microparticles (electrons, photons, protons, etc.) cannot have simultaneously determined positions and velocities: the more accurately the position is determined, the larger is the uncertainty of the velocity, and vice versa. A similar uncertainty principle (W. Heisenberg, 1927) holds for the energy and (life)time of excited quantum systems. Thus the complete description proved impossible for microscopic systems. (The developments in nonlinear dynamics and deterministic chaos in the last decades of 20th century demonstrated that complete description is a rare bird in macroscopic systems too.)

Turning back to microbiology, we note that each step down the scale of description leads to a more mechanical and simpler behavior. The bacterium is more mechanical than the drosophila, the DNA molecule is more mechanical than the bacterium.

Quantum theory, however, has shown that further steps down the scales have exactly the opposite effect. The atoms constituting DNA have less expressed mechanical behaviour than the molecule itself, and the nucleus and the electrons constituting the atoms, are less mechanical than the atoms. Classical behavior is being replaced by the quantum mechanical one, with its typical uncertainty.

The less mechanical the behavior of microparticles the more essential becomes the role of the observer. The view of an electron existing in an objective state that is independent of the observer is no longer valid. The part of the observer is interpreted in different ways but it is an undeniable experimental fact that the description of subatomic processes depends on the method of observation. On dealing with micro-objects the observer cannot be excluded from the description of Nature. The laws of microphysics cannot even be formulated without a reference to the observer (the subject of the cognitive process) and to his mind (consciousness). In such a way physics discovered within the laws of (inanimate) Nature a specific niche for human mind, the subject and his consciousness (Dyson, *Disturbing*).

Presently, we are eyewitnesses to a specific ironical situation in modern science. Biological sciences that previously postulated the privileged role of human mind in nature's hierarchy, today, following the path of the 19th century physics, practically exclude mind from their scope. At the same time physics, faced with compelling experimental evidence, has been moving away from a strictly mechanistic picture of the Universe and advances to a view in which mind has been given an integrating role in the physical reality.

> "It is as if", remarked the psychologist H. Morowitz (quoted after P. Davies, *opus cit.*), "the two disciplines were on fast moving trains, going in opposite directions and not noticing what is happening across the tracks."

It is this paradoxical situation, as stressed by Prigogine and Stengers, that outlines the context in which today's metamorphosis of science proceeds. In its hard core this metamorphosis is based upon the contradiction between the view, on one side, of the creative role of man in science and, on the other side, the classical science view of man as a dispassionate subject of cognition, isolated from nature, acting behind the screen, excluded from the context of Nature's evolution.

7.

Quite remarkably, mind appears on three different levels in the physical description of nature.

On the level of microphysics the observer is inseparably linked to the objects of observation (for more details see part **8**). On the macrolevel, i.e., the level of the immediate life experience, we are aware of our own mind, we rationalize the surrounding world (in particular, we discuss the role of human agency in it), we have feelings and we give ethical and aesthetic evaluations. This is the level of psychology and of the other humanities, as well as of the arts.

In between these two levels is the level of microbiology in which the mechanical models "work" but mind is not a part of the processes.

Finally, on the megalevel the validity of the anthropic principle is assumed, to wit, of the hypothesis of a specific type of conformity between the structure and the functioning of the Universe, on the one hand, and the existence of man and of human mind, on the other. As we are to see (part **9**), there is a kind of extremely fine "tuning" among the fundamental characteristics of the physical world, and this favors in an extraordinary manner the appearance and the existence of intelligent life.

8.

In quantum mechanics the assumed objectivity of subatomic particles and processes proves to be too crude a simplification of reality. If one wishes to gain certain notions about the behavior of microparticles, it is principally impossible to ignore the physical processes by which the information about them is obtained. In the classical world the processes of measurement have a subordinate role (though even there this is not always true: classical chaotic systems are characterized by an extremely strong sensitivity to the slightest changes in external conditions - the so-called "butterfly effect"). However, with microparticles any act of observation causes serious perturbations. Thus we can no longer speak of the behavior of a particle regardless of its dependence on the process of observation. Unless we have a full description of the measuring apparatus, we should not even consider micro-objects as *objects of definite properties,* said Niels Bohr. And Werner Heisenberg (Heisenberg, *Das Naturbild*) added firmly:

> As a result of this the natural laws have already no direct bearing on the elementary particles themselves, but on our knowledge about them.

In quantum physics one can make statements about what is happening when the particle, whose behavior is being recorded, interacts with another physical system, say, the measuring device. When there is a quantum measurement the interaction cannot be regarded as a unidirectional action of the object upon the subject, because the measurement changes not only the state of the device but of the object too. The theory of quantum measurement revealed the activity of the cognitive subject, as well as the complexity of the relation object - subject.

The physics of microphenomena is only one link in the endless chain of the man - nature interrelations; however, this science cannot make statements about Nature "as it is" because

> "Cognition of nature", says Heisenberg, "already assumes invariably **the human presence."** (Heisenberg, *opus cit.*;my italics; M.B.)

We ought to acknowledge the fact that *we are not only spectators but we are at the same time actors in the great drama of existence* (N. Bohr).

In quantum theory, wrote Heisenberg (Heisenberg, *opus cit.*), we find ourselves from the very beginning on the intersection of the interrelations between nature and man.

> Thus, the universally adopted common division of the world into object and subject, inner world and outer world, body and soul is no longer adequate.

Consequently, in natural sciences the object of study is already not nature in itself, but nature in which "*man rediscovers himself*" (Heisenberg, *opus cit.*).

The uncertainty principle makes meaningless questions of the type: "where is the electron positioned and how is it moving"? Only observation lends particular reality to the microparticles and the observer is the one to decide what is specifically to be observed - coordinates or velocities or other characteristics of microparticles.

The idea that observations create the reality in microworld is openly blasphemous to the classical way of thinking. "Isn't it true", asked Einstein (A. Einstein, *Überlegungen*) emphatically in one of his latest papers on the interpretation of quantum mechanics, "that the real world exists independently, no matter whether we observe it or not."

In 1935 Einstein, jointly with B. Podolsky and N. Rosen, suggested a thought experiment which, hopefully, would make clear this question. The idea was by one measurement to obtain the exact value of the location of a particle at a given moment, and by another measurement at the same moment to determine the exact value of its velocity. Thus the expectation of Einstein would be justified that reality in quantum mechanics is objectively determined. Thanks to the theorem of J. Bell (1964), the EPR thought experiment could be accomplished in real laboratory conditions (A. Aspect et al., 1982). It proved that the reality is not objectively determined but is really and truly *created by the process of observation.* The dictated by the "common sense" view of the world consisting of objects that exist "in themselves", independently of our observations, completely loses its meaning in the quantum world.

Many of the strange ("ghost" as L. Lederman put it) peculiarities of the microworld boil down to the wave-particle duality, which somehow resembles the mind-body duality. In both cases these are two dualistic aspects of one and the same entity: they preclude each other and, at the same time complement each other. The physical reality is not the sum of these two alternatives but rather a flexible and interpenetrating combination - a *hybrid reality* (P. Davies, *opus cit.*). Each one of our (experimental) interventions determines fundamentally the character of physical reality. This is emphatically demonstrated by the phenomenon of wave-function reduction (collapse). Until the microobject is described by a wave it can be found everywhere in space. However, when the measurement has been performed the wave reduces to a single state. John von Neumann proved (Von Neumann J., *Mathematische*) that such an instantaneous "collapse" comprises also the large system of the measuring device, thus recognizing grounds to refer to the psychophysical parallelism (the relationship between psychic and physical phenomena) of Leibnitz. This inference was ridiculed by Schrödinger (1935) by proposing the renowned paradox of the cat, which, being considered as a quantum hybrid, can exist in the absurd superposition of "live" and "dead" states (until moment is observed by someone or by something and fixes itself in one of the two states). E. Wigner (1962) daringly raised the question that for a long time disturbed many people: if body acts upon mind, then why shoudn't mind act (action and reaction) upon body? Assuming this to be the case, then wave-function collapse is the result of the participation of a conscious observer in the long chain of the observation equipment.

Something more, as pointed out ironically by J. Wheeler (1979, Princeton symposium on the occasion of Einstein's anniversary), we can even *delay the choice* of the observational strategy until the hybrid world has already come into being. In such a way, claimed Wheeler, mind can be made responsible for the retroactive creation of reality - even of the reality that existed before there were people (quoted after P. Davies, *opus cit.*).

This illustrates and supports the thought of Heisenberg that there is already no point in the common division of the world into subject and object, inner world and outer world, body and soul. This is the reason Paul Davies (Davies, *opus cit.*) made the emphatic conclusion that

> Quantum theory provides the most convincing scientific evidence yet that consciousness plays an essential role in the nature of physical reality.
>
> "What is to be done", exclaimed the late Bulgarian philosopher of science Sava Petrov (Petrov A. and S. Petrov, *Quantum*), "that the decision of the subject can determine how a given object will realize itself - as a wave or as a particle? And, to make matters worse, the decision to put in operation one or another device can most likely be delayed to such a degree that it determines not only the future, but also the past of the micro-object which is experimented on."

We could continue in the same spirit: what is also to be done with the quantum Zeno effect (Khalfin, *Quantum*; Misra and Sudarshan, *Zeno's*) in which, paradoxically, the interaction of the measuring device with an unstable microsystem can delay and even bring to a halt its decay? Still, the effect has really been observed for the transitions of excited atoms. This would indeed be, as stressed by I. Prigogine et al., an important step towards the understanding of the role of the observer in quantum mechanics (Petroski et al., *Quantum*)

This appearance of mind in the arena of the physical world is quite stunning.[1] That is why N.Bohr, as early as in 1937, sagaciously cautioned:

> such inferences by no means connote that in atomic physics mysticism of any kind is tolerated.

Unfortunately, the attempts to mystify quantum physics are not so scanty (see e.g., the book by L. Lederman and D. Teresi "The God Particle", Interlude B "*The dancing masters of moo-shu*"and the book by M. Gell-Mann "*The Quark and the Jaguar*", ch. 12 "*Quantum mechanics and nonsense*").

Nevertheless, one way or another, the anthropic agent manifests itself on the microlevel and offers, precisely to physics, serious reasons to go on a quest for the solution of the problem of mind (see the book by R. Penrose, *opus cit.* and the first one on the same topic: Penrose, *The Emperor's*).

9.

Why did the physical world originate with the known fundamental properties: tridimensional space, unidimensional time, space-time continuum and specific numerical values of the fundamental physical constants? Such questions sound quite bizarre even to modern physics as these physical properties of the world, as a rule, are postulated. In addition, questions of the type "*why?*" and "*for what purpose?*" seem to revert us to the vitalistic physics of Aristotle with its *causes finales*.

Still, P. Ehrenfest, as early as in 1917, in a half-forgotten work offered an answer for classical systems: space is tridimensional *because* in such only space stable structures like the planet systems can exist (later on the same proposition was also proved for quantum-mechanical systems).

The whole set of fundamental physical constants is so delicately matched that even extremely small changes in it would destroy (or would frustrate the formation of) basic stable structures (atomic nuclei, atoms, stars, galaxies).

Thus, for instance, if in the first second after the Big Bang the density of the Universe were bigger even by one trillionth part, the Universe would have collapsed in 10 years. On the other hand, if the density of the Universe were by the same amount less, then it would be almost empty (Hawking, *Black Holes*).

The attractive forces between the nucleons in atomic nuclei are exactly as big as necessary in order to counterbalance the electric forces of repulsion between the positively charged protons. However, the nuclear forces are not large enough to bind two protons into one nucleus - diproton. If the diproton existed in nature almost all the hydrogen would have bound heavier nuclei. Hydrogen itself would then be a rare element and stars similar to our Sun, whose main life cycle is based upon the burning of hydrogen in their cores, would not exist. While, on the other hand, if the nuclear forces were considerably weaker than the real ones, hydrogen would not be able to burn and there would be no heavy elements. Consequently, if the evolution of life needs a star like our Sun producing energy at a constant rate for millions of years, then, in order that life would be possible, the magnitude of the nuclear forces should come into a very narrow range.

If the electron mass was increased over 2.5 times, the proton and the electron in the hydrogen atom would convert into neutron and neutrino, so that the Universe would consist only of neutron stars.

If the carbon isotope C^{12} did not have an energetic level exactly equal to 7.82 *MeV*, the chemical elements of the periodic table would not have formed. The substance of the Universe exists owing to this incredible coincidence (see the book by R. Feynman, *The Character*; however, recently S. Weinberg demonstrated that this restriction is not so precise with a tolerance of about 20%; S. Weinberg, *A Designer*).

It is difficult even to enumerate all the "lucky" coincidences in the fundamental structure and functioning of the physical world, that represent the prerequisites for the existence of life and conscience. Without such coincidences water could not exist as a liquid, the chains of carbon atoms would not form the complex organic molecules, and hydrogen atoms would not establish bonds between them (F. Dyson, *opus cit.*). The impression emerges of an extremely fine "tuning" of physical constants and laws, which makes the Universe unexpectedly hospitable for the existence of living beings and especially of human beings.

How could one explain the coincidences of the big numbers (A. Eddington, P. Dirac, 1937), that is, the observation that the dimensionless combinations of fundamental physical constants (Planck constant, light speed, gravitational constant, electron mass and charge, etc.) always have in a mysterious way values of the order of the formidable number 10^{40}? How could one explain the countless "favorable" coincidences that determine the existence of life and of consciousness?

First R. Dicke in the 1960's (followed by S. Hawking, B. Carter, J. Barrow, G. Whitrow and many others) made an attempt to explain these coincidences with the help of a hypothesis that later became known as the *anthropic principle.* According to this, the very fact that there exists an observer and the fact that this observer has a natural (not divine) origin, sets heavy restrictions on the structure and the evolution of the Universe. According to a definition of the anthropic principle given by S. Hawking (Hawking, *A Brief*):

> we see the universe the way it is because if it were different, we would not be here to observe it.

It is the central role of man in the physical world that is emphasized in aphoristic definitions of the anthropic principle:

> Here is man, what the Universe should be? (J. Wheeler)

and

> Cogito, ergo mundus talis est (Brandon Carter's variation of Blaise Pascal's famous phrase).

The anthropic principle is a kind of return to geocentrism. This time, however, the Earth is not given a privileged position as a physical body - a planet (as such it is too trivial), but as the cradle (perhaps unique) of human civilization.

The anthropic principle (for many more details see the fundamental monograph: by Barrow and Tipler, *The Anthropic*) came up to expectations by a number of important achievements in the interpretation of certain cosmological relations, for example, between the age of the Universe and certain fundamental physical constants, the coincidence of the big numbers, the relation between the average density of the Universe and critical density (the latter conditioning the geometry of the Universe - whether open, flat or closed), and many others. In all such cases the argumentation boils down in the final analysis to the requirement for the fulfillment of one or another necessary condition for the existence of conscious life.

The arguments of the anthropic principle, being focused at the role of man in the Universe, are unconditionally useful against the absence of a deeper physical principle. (As B. Shaw had put it:

> The hen is a means for the egg to create another egg.

Isn't the anthropic principle trying to tell us that "the Universe is a means for man to create another man"?) It may well be that the anthropic principle will in the last reckoning prove tautological, an expression of that paradoxical self-reference, which Douglas Hofstadter called "*a strange loop*" (Hofstadter, *Gödel*) and Maurits Escher pictured brilliantly in his lithograph "*Drawing hands*". It is quite likely that life is the result of self-organization processes and that consciousness is a natural outcome (not a *causa finale*) of "*selective advantages*" (R. Penrose, *The Emperor's*).

Yet it is quite symptomatic that contemporary physics is inclined to attribute universal cosmic meaning to the origin and existence of man. In this way the human agency attains a relative independence, its roots penetrate deeply into the fundamental properties of the Universe, and it transforms into an active regulator of physical phenomena.

10.

Contemporary science reveals the importance of human agency on three clearly discernible levels. The question remains open, however, of the connection between these three levels of manifestation of human consciousness.

Roger Penrose made a daring attempt (Penrose, *The Emperor's*) to tie together the physical effect of collapse of the wave function, which baffles formalization, with the non-algorithmic nature of human mind and thus to unify micro- and macro-levels.

The probabilistic behavior of quantum objects could be one of the premises of the free and unpredictable behavior of individuals. The puzzling features of "becoming aware" of a situation and of "intuitive dawning" with their discrete and unpredictable character are strongly reminiscent of the process of wave-function reduction. John von Neumann (*opus cit.*) claims that the act of reduction (collapse) cannot be understood without addressing the abstract subject, namely, without making clear the still unknown connection between quantum physics and psychology (probably, some kind of a unified consciousness). Eugene Wigner assumed (*opus cit.*) that it is sufficient to reformulate some of the fundamental physical laws and with their help it would become possible to describe the properties of life and consciousness. E. Schrödinger (1947) and later Manfred Eigen (1966) and Ilya Prigogine (1973) have shown that life really lends itself to physical description.

> Thus more and more scientists become convinced in the prospects opening before the interconnection of quantum physics and psychology (Grib, *Quantum Logic*).

Jacques Monod warns that "the ancient connection" has irretrievably vanished. This is so, but at the same time the old Descartes' division of the world into *res extensa* and *res cogitans*, i.e. the external world of the objective events in space and time, and the consciousness that reflects these events, is also unacceptable to modern physics. Science does not take the position of a passive observer, it becomes aware of itself as a specific type of interaction between man and Nature, so that

> man finds himself facing his own nature (W. Heisenberg, *opus cit.*).

The reply to Monod's warning was given by Prigogine – with the book bearing the significant title "*The New Alliance*" (Prigogine, Stengers, *opus cit.*). It is stressed that we are not to quest for a "reanimation" of animism but for radically new concepts, which "place man within the world described by him" (*opus cit.*).

It is this conceptual change that Prigogine emphatically calls "*metamorphosis of science*".

As a result of this metamorphosis the picture of the physical world is no longer the one given by natural science but also involves anthropic factors. It is quite possible that in the future physics will suggest to the humanities a new model of scientific research, a model that differs from the classical one and takes into account the inextricable connection object – subject, the one that Goethe's Mephistopheles calls "*das Geistigeband*" – the spiritual link.

Man is the measure of all things.

On pronouncing this sacral phrase did Protagoras from Abdera imagine that 25 centuries later science would relate the atomism of his contemporary and fellow-citizen Democritus to human nature?

NOTES

[1] The well-versed reader might be surprised at the observation that, having a wealth of interpretations of quantum mechanics (see, e.g., the review of S.Sonego, *Conceptual*), among which the attempts to save objectivism (such as the idea of "many alternative worlds"), here the stress is laid on those that highlight the role of the observer. This, however, is precisely our aim here (see the introductory paragraphs).

REFERENCES

Barrow, J. and F.Tipler. *The Anthropic Cosmological Principle.* Oxford Univ.Press, Oxford,1986.

Davies, Paul. *God and the New Physics.* Penguin, 1983

Dyson, Freeman. *Disturbing the Universe.* Harper & Row, 1979.

Einstein, Albert. "Überlegungen zur Interpretation der Grundlagen der Quanten-Mechanik. In: *Scientific Papers, presented to Max Born.* Edinburgh: Oliver & Boyd, 1953

Feynman, Richard. *The Character of Physical Law.* Cox & Wyman, London, 1965 (lecture 5).

Gell-Mann, Murray. *The Quark and the Jaguar.* Little, Brown & Co, London, 1994.

Grib, A.*Quantum Logic: possible directions* (in Russian). Moscow, 1987.

Hawking, Stephen. *Black Holes and Baby Universes.* Bantam Books, New York et al., 1993 (ch. 13).

Hawking, Stephen. *A Brief History of Time.* Bantam Books, London et al., 1988 (ch. 8).

Heisenberg, Werner. *Das Naturbild der heutigen Physik.* Hamburg, 1955.

Hofstadter, Douglas. *Gödel, Escher, Bach: An Eternal Golden Braid.* Penguin, New York, 1981.

Khalfin, L.A. "Quantum Theory of the Decay of Physical Systems (in Russian). *JETP* **33** (1958) 1371; Misra, B. and E.Sudarshan. "The Zeno's Paradox in Quantum Theory". *J.Math.Phys.*, **18** (1977) 756.

Lederman, Leon and Dick Teresi. *The God Particle: if the universe is the answer, what is the question?.* NY, 1993.

Monod, Jacques. *Le hasard et la necessité.* Seuil, 1970.

Penrose, Roger. *The Emperor's New Mind: Concerning Computers, Minds and the Laws of Physics.* Oxford Univ.Press, N.Y., 1989.

Penrose, Roger. *Shadows of the Mind: a Search for the missing Science of Consciousness.*Oxford Univ.Press, NY, 1994.

Petrosky, T.,S.Tasaki and I.Prigogine. "Quantum Zeno Effect". *Physica* A **170** (1991) 306.

Petrov, Assen and Sava Petrov, *Quantum Mechanics* (in Bulgarian). Sofia, 1989.

Prigogine, Ilya et Isabelle Stengers. *La Nouvelle Alliance: Métamorphose de la Science.* Gallimard, Paris, 1986.

Russell, Bertrand. *Science and Religion.* London, 1935.

Sonego, S. "Conceptual foundations of quantum theory: a map of the land". *Annales de la Fondation Louis de Broglie*, **17** (1992) 405.

Von Neumann, John. *Mathematische Grundlagen der Quantenmechanik.* Springer, Berlin, 1932; English translation: *Mathematical Foundations of Quantum Mechanics.* Princeton, 1955.
Weinberg, Steven. *The first three Minutes.* NY, 1977.
Weinberg, Steven. *A Designer Universe?*, The New York Review, October 21, 1999.
Wigner, Eugene. "Remarks on the Mind-Body Question". In *The Scientist Speculates*, Heinemann, London, 1961.See also: *Am.J.Phys.* **31** (1963) 6.

Part III

PHILOSOPHY AND LOGIC

VLADIMIR SOTIROV

LEIBNIZ'S LOGICAL SYSTEMS: A RECONSTRUCTION

1. INTRODUCTORY REMARKS

The system of logic plays a central rôle in Leibniz's methodology of science. According to his basic ideas, any scientific question should be posed in the form "Is *S* a *P*?" and its answer might be found either in *Encyclopædia Universalis* of the human knowledge (when the fact is synthetic) or in the corresponding relation between *Numeri Characterici* of *S* and *P* (when the fact is analytic). Leibniz believed that "a few selected men", as he wrote, could systematize (in a *quinquenium*!) all data and establish a correspondence between the elementary notions and their characteristic numbers. As we see, the logical system determines both the formal translation of the scientific sentences and the mechanism of checking their truthfulness.

In the style of his epoch, Leibniz has identified the whole logical reasoning with syllogistic. His efforts to turn human reasoning into some kind of mathematics have been in two directions. The first one may be called *algebraic*. It is presented in various manuscripts during all his life. The main goal has been the abstract representation of the syllogistic laws and in such a way, obtaining a special algebra of notions. The second direction of investigations (all of them concentrated in April 1679) may be called *arithmetical*. The central idea has been embedding that logical algebra into arithmetic. In other words, to establish an isomorphism between the logical algebra of terms and a suitable algebra of integers. Therefore, Leibniz's "Calculemus!" was meant to construct an adequate model of the basic syllogistic relations and operations into the arithmetic of integers.

All Leibniz's trials to find some natural model of syllogistic can be retraced from the first publications of his logical papers (Couturat, *Logique* and *Opuscules*) as well as from the new academic edition of his philosophical manuscripts (Leibniz, *Sämtliche Schriften*). Unfortunately, those trials were unsuccessful. Finally Leibniz has radically changed the arithmetical interpretation: in the place of simple relations between integers based on divisibility, relations between *pairs* of (relatively prime) numbers appeared. In 1946, J. Słupecki (*Syllogistic*) proved that the last model was adequate; see also the celebrated book of J. Łukasiewicz (*Aristotle's Syllogistic*). However, the model of pairs possesses two big disadvantages. The first one is that the initial naturalness has died. Actually, it is not fatal from the mathematical point of view. However, the second disadvantage is principal: it is not clear how the

Dimitri Ginev (ed.), Bulgarian Studies in the Philosophy of Science, 85-93.

model of pairs of integers can be extended to involve the generalizations of the Aristotelian syllogistic containing, e.g., term negation or term conjunction.

It was shown (Sotirov, *Arithmetizations*) that Leibniz's primary plan to explore divisibility of integers had been vital and two kinds of arithmetical models adequate to many systems of syllogistic were built up. Later on the arithmetical interpretation was extended to cover the pure monadic predicate calculus, i.e., the entire logic of properties (Sotirov, *Monadic*). Finally, it was shown that the Leibniz style arithmetization could be extended to the monadic predicate calculus containing equality as well (Sotirov, *Equality*). All papers cited here are available from *http://www.math.bas.bg/~vlsot*; the detailed proofs can also be found there.

The following methodological notice may be added here: any variant of syllogistic is decidable while the whole arithmetic is not. In such a way, speaking *in abstracto*, the calculation is not able to increase the "algorithmicity" of any disputation according to Leibniz's dreams. Nevertheless, that fragment of arithmetic in which all variants of syllogistic were imbedded is decidable because it contains *multiplication* only (Skolem, *Satzfunktionen*; Boolos and Jeffrey, *Computability*, § 25). The same fragment was used for the monadic predicate calculus (with or without equality).

The present paper is neither historical, philosophical nor archival. Its goal is not to expose the development of Leibniz's ideas, their philosophical underground or their exact places in his manuscripts. Our aim is to reconstruct the variety of logical systems sketched by Leibniz and in such a way, to obtain a working mechanism: probably the mechanism that Leibniz would describe if he was alive today.

2. ARITHMETIZATIONS OF THE TRADITIONAL SYLLOGISTIC

The Aristotelian syllogistic will be treated in the style that became canonical (Łukasiewicz, *Aristotle's Syllogistic*). For this purpose the language of the classical propositional calculus is extended by *term variables* or, for short, *terms* t_1, t_2,... together with two binary *term relations*: A and I. Syllogistic *atoms* are all formulae of the kind $s\mathrm{A}p$ or $s\mathrm{I}p$ with s and p being terms. A *syllogism* is any propositional formula with all propositional letters replaced by syllogistic atoms. The formal connectives used are $\neg$, $\wedge$, $\vee$, $\Rightarrow$; &, $\rightarrow$ and $\leftrightarrow$ are used as informal abbreviations.

The standard and the most intuitive semantics of the Aristotelian syllogistic is that in the theory of sets: if S and P are arbitrary *non-empty* sets, $s\mathrm{A}p$ ("Every s is a p") is translated into $S \subseteq P$, $s\mathrm{I}p$ ("Some s is a p") into $S \cap P \neq \varnothing$, and the formal propositional connectives are replaced by the informal ones. Thus any syllogism is translated into a sentence about non-empty sets. If this sentence is true, i.e., if the expression so obtained is a set-theoretical tautology, the syllogism is said to be *true*. It is *true in a given* (*non-empty*) *set* U when any replacement of its terms with (non-empty) subsets of U gives a true sentence. This semantics will be called *Scholastic* according to Leibniz's own use of this word.

Another semantics in the theory of sets is possible; it will be named *Leibnizian*, being (partially) accepted by him. When a non-empty set U is given, term variables are evaluated by subsets of U *different from* U. If S and P are such sets, $s\mathrm{A}p$ is interpreted as $S \supseteq P$, $s\mathrm{I}p$ as $S \cup P \neq \mathrm{U}$, and the formal propositional connectives are replaced by informal ones. A syllogism is said to be *true in* U when the sentence obtained after any replacement of all term variables with subsets of U (different from U) is true. The syllogism is *true* when it is true in any set U.

On the base of both set-theoretical intuitions, two translations of the traditional syllogistic into arithmetic are possible. The first one will be named *Scholastic*. Let $a_1, a_2, \ldots$ denote arbitrary integers *greater than* 1. Given a syllogism, replace $t_i\mathrm{A}t_j$ with "a_i is a divisor of a_j", $t_i\mathrm{I}t_j$ with "a_i and a_j have a common divisor greater than 1", or: g.c.d. $(a_i, a_j) > 1$, and the formal propositional connectives with informal ones. Call the syllogism *arithmetically true* (in the *Scholastic sense*) if the sentence so obtained is an arithmetical truth.

Fact 1 (Adequacy of the Scholastic arithmetical interpretation): *A traditional syllogism is true if and only if it is arithmetically true (in the Scholastic sense).*

If the empty set is admitted to evaluate terms in the Scholastic semantics, the arithmetical interpretation can be modified: 1 has to be added to the list of divisors, being the number corresponding to empty terms.

Further, for the arithmetical interpretation named *Leibnizian*, let u (the Universe number) be an arbitrary integer greater than 1, and let $a_1, a_2, \ldots$ be its arbitrary proper divisors, i.e., $a_i < u$ for any i (however, $a_i = 1$ is permitted). Replace $t_i\mathrm{A}t_j$ with a relation "a_i is divisible by a_j", and $t_i\mathrm{I}t_j$ with "there is a *prime* divisor of u dividing neither a_i nor a_j", or: l.c.m. $(a_i, a_j) < u$. Finally, formal propositional connectives are replaced with their informal analogues. The syllogism is said to be *arithmetically true* (in the *Leibnizian sense*) *with respect to u* if the sentence so obtained is an arithmetical truth. The syllogism is *arithmetically true* (in the *Leibnizian sense*) if it is arithmetically true, in the same sense, with respect to any $u > 1$.

Fact 2 (Adequacy of the Leibnizian arithmetical interpretation): *A traditional syllogism is true if and only if it is arithmetically true in the Leibnizian sense.*

In this semantics, if empty terms are admitted, they have to be evaluated by the Universe number u.

3. ARITHMETIZATIONS OF SYLLOGISTIC WITH NEGATIVE TERMS

Expand the language of syllogistic by adding an operation of term negation –; then, if t is a term, $-t$ ("non-t") is a term, too. The definition of atoms is modified by permitting s and p to be arbitrary terms in $s\mathrm{A}p$ and in $s\mathrm{I}p$ as well. In both set-theoretical semantics, a *universal set U* is introduced. According to the tradition, terms are evaluated by subsets of U different from $\varnothing$ and U. If t is evaluated by a set T, the value of $-t$ is the complement of T to U. The rest of the definition of a true syllogism remains the same. In both arithmetical interpretations, a Universe number $u > 1$ *without multiple factors* is introduced together with the following rules: 1) all evaluating integers are divisors of u different from 1 and u; 2) if term t is evaluated by an integer a then the term $-t$ is evaluated by the fraction u/a.

Fact 3 (Adequacy of both Scholastic and Leibnizian arithmetical interpretations of syllogistic with term negation): *A syllogism (possibly with* *negative terms) is true if and only if it is arithmetically true in the Scholastic as well as in the Leibnizian sense.*

4. ARITHMETIZATIONS OF SYLLOGISTIC WITH TERM COMPOSITION

In this section neither empty nor universal terms will be rejected. The treatment of the term composition may be made independent of the presence of term negation. However, if negation does occur together with a composition, all Boolean term operations will be defined. That is why it will be better to consider a full Boolean algebra straight away.

Following Leibniz, the composition of any terms *s* and *p* will be noted by *sp*. The class of terms now is the smallest class involving term variables and closed under negation and composition. Given a Universe $U \neq \varnothing$, an *evaluation* of a term *t* in *U* is a set *T* obtained after replacing all term variables in *t* with subsets of *U* as well as term operations with their corresponding set-theoretical operations. Namely, the composition is interpreted as an *intersection* in the Scholastic semantics, and as a *union* in the Leibnizian one. Having terms evaluated, the translation of a syllogism into a set-theoretical sentence remains the same as in Section 2.

Further, in both arithmetical interpretations, term variables will be evaluated by arbitrary divisors of a Universe number $u > 1$ *without multiple factors*. The evaluation of the negation remains the same. If terms s_1 and s_2 are evaluated by integers *a* and *b*, their composition s_1s_2 will be modeled by g.c.d. (*a*, *b*) in the Scholastic arithmetical interpretation, and by l.c.m. (*a*, *b*) in the Leibnizian one.

Fact 4 (Adequacy of both Scholastic and Leibnizian arithmetical interpretations of syllogistic with all Boolean term operations): *A syllogism (possibly containing arbitrary Boolean term operations) is true if and only if it is arithmetically true in the Scholastic as well as in the Leibnizian sense.*

Let us note that perhaps E. Bunitsky (*Applications*) was the first to use in 1899 the Boolean algebra consisting of all divisors of a given integer without multiple factors.

5. ARITHMETIZATION OF THE PURE MONADIC PREDICATE CALCULUS

The language of the monadic calculus contains *individual variables x, y, z,...*, one-place *predicate symbols* P_1, P_2,..., quantifiers ∀ and ∃, and the usual propositional connectives with brackets. A *monadic proposition* is a formula without free variables. For such formulae we adopt the following result (Boolos and Jeffrey, *Computability*, Theorem 25.4):

Lemma: *Any monadic proposition is equivalent to a monadic proposition with the same predicate symbols and one variable only.*

Let the sole variable be *x*. In addition, we may suppose it is not bound "twice" anywhere. So, no formula under consideration is a Boolean combination of two subformulae *A*(*x*) and *B*, one of them containing a *free x* and the other containing *x bound*; in (Q*x*)*A*(*x*), where Q is a quantificator, a free *x* does occur in *A*.

To build up an arithmetical model for monadic propositions, let an arbitrary integer $u > 1$ *without multiple factors* be taken, and let its divisor d_i be associated with the predicate $P_i(x)$. Following the construction of the formula, a *divisor* of u will be associated with any subformula containing a free x, and a *statement* about divisors will be associated with the subformula when it does not contain a free x. If a and b are associated with $A(x)$ and $B(x)$, respectively, then g.c.d. (a, b) is associated with $A(x) \wedge B(x)$, u/a with $\neg A(x)$, and so on for other Boolean connectives; the statements $a = u$ and $a > 1$ are associated with $(\forall x)A(x)$ and $(\exists x)A(x)$, respectively; if statements p and q are associated with subformulae A and B, then "p and q" and "not p" will be associated with $A \wedge B$ and $\neg A$, respectively. Finally, a certain statement comparing divisors of u with u and 1 will model the initial monadic proposition. If this statement is an arithmetical truth for an arbitrary integer u, the proposition is called *arithmetically true*. Using that any predicate tautology is equivalent to a closed formula, we obtain the main

Fact 5: *Any monadic formula is a predicate tautology if and only if its corresponding monadic proposition is arithmetically true.*

There is a well-known translation of both basic syllogistic relations into the pure monadic predicate calculus: $s\text{A}p$ is $(\forall x)(S(x) \Rightarrow P(x))$ and $s\text{I}p$ is $(\exists x)(S(x) \wedge P(x))$. Applied to the predicate formulas just written, the arithmetical interpretation described above will produce the Scholastic arithmetical semantics of syllogistic. The dual (i. e., the Leibnizian) semantics of syllogistic could be extended to the whole monadic predicate calculus as well (with or without equality) but then the interpretation of any formula would simply coincide with the Scholastic interpretation of *the negation of its dual*.

6. ARITHMETIZATION OF THE MONADIC PREDICATE CALCULUS WITH EQUALITY

Now, the language of the pure monadic predicate calculus is extended by the only dyadic predicate =. Formulas are defined in the usual way. For this language, *models* in non-empty domains and *evaluations* of individual variables in them are introduced in the standard manner (Kleene, *Logic*, § 29].

For the *arithmetical* models, let $N > 1$ be an integer *without multiple factors*. Any predicate P_i is interpreted by arbitrary divisor of N (possibly 1 or N) denoted with $d(P_i)$, and any individual variable x_i is evaluated by $d(x_i)$, a *prime* divisor of N. Following the construction of a formula F, its *arithmetical statement* AR[F] corresponding to given evaluation will be obtained. Namely, for atomic formulas, AR[$P_i(x_j)$] is "$d(x_j)$ divides $d(P_i)$" and AR[x_i=x_j] is "$d(x_i) = d(x_j)$"; for a subformula G, AR[$(\forall x)G$] is "for any prime divisor d, $\text{AR}^x_d[G]$" where AR^x_d differs from AR attaching d to x. Finally, all propositional connectives are replaced with their non-formal analogues. If AR[F] is a true arithmetical sentence for any N under arbitrary evaluation, F is named *arithmetically true*. This semantics is relevant:

Fact 6: *Any formula of the monadic predicate calculus with equality is a tautology if and only if it is arithmetically true.*

7. VARIOUS SYLLOGISTICS FROM THE ALGEBRAIC POINT OF VIEW

Let us begin with the *traditional* (*Aristotelian*) *syllogistic*. As we know from the result of Słupecki (*Syllogistic*), the syllogisms not admitting empty terms are axiomatized by the following four axioms of Łukasiewicz: *s*A*s*, *s*I*s*, (*m*A*p*)∧(*s*A*m*) ⇒ (*s*A*p*) (*Barbara*), and (*m*A*p*)∧(*m*I*s*) ⇒ (*s*I*p*) (*Datisi*).

Let us consider A and I as binary relations $\leq$ and θ in an arbitrary class of objects, and take ∧, ⇒ in their informal sense. Then *s*A*s* and *Barbara* show that $\leq$ is *reflexive* and *transitive*, i. e., it is a *quasi-ordering relation*. To obtain an algebra, $\leq$ has to be *anti-symmetric*, i. e., an additional axiom (*s*A*p*)∧(*p*A*s*) ⇒ ($s = p$) is needed. Strictly speaking, = here belongs to the extended language of syllogistic but it is not difficult to carry out such extension formally. This innovation was made for first time by Leibniz; see, e. g., *Elementa Calculi* (Couturat, Opuscules, 52). In such a way, the set of terms turns into a *partially ordered structure* (Birkhoff, *Lattice*). The second relation θ is *reflexive* and *monotonic* with respect to $\leq$. So the traditional syllogistic may be shortly characterized by a pair ($\leq$, θ).

Actually, in most of his logical manuscripts, Leibniz introduces into the syllogistic a *composition* of terms besides A and I. We prefer this neutral name (adopted from Leibniz himself) in order to avoid any specifications of its "real" nature: is it a term conjunction or a term disjunction. Then some transpositions in the expressive means of the syllogistic language become possible. Let us denote by ° the algebraic analogue of the composition. Irrespective of what it is in the semantics, an intersection or a union, as a minimum it should be *idempotent*, *commutative*, and *associative*. To be formulated, these three properties require only the term *equality* to appear explicitly. A structure with an operation as described G. Birkhoff names *semi-lattice*. An ordering relation $\leq$ we need in syllogistic may be defined in such a structure by $x \leq y$ *iff* $x = x°y$. Indeed, its *reflexivity* follows from the idempotency of °, its *transitivity* from the associativity, and its *anti-symmetry* from the commutativity. On the other hand, with respect to $\leq$ so introduced, the composition has the properties of *inf*: $x°y \leq x$; $x°y \leq y$; if $x \leq y$ and $x \leq z$ then $x \leq y°z$. These obvious results about the system (=, °) are summarized in the following

Fact 7: *Any operation ° which is idempotent, commutative and associative produces an ordering relation $\leq$ which is reflexive, transitive and anti-symmetric; moreover, $x°y = inf(x, y)$ with respect to $\leq$. A second ordering may be defined by $x \leq y \leftrightarrow y = x°y$; the composition will be then a sup with respect $\leq$ so defined.*

The possibility to build up syllogistic in this way has been noted many times by Leibniz. In fact, he does list idempotency and commutativity of the composition, defines A as it was shown; see, e. g., *Primaria Calculi Logici fundamenta* (Couturat, Opuscules, 235–236). Only associativity was not mentioned explicitly; however, it is described in a significant example (Couturat, Opuscules, 258).

It would be natural to have *two* dual operations simultaneously, ° and •, one of them playing the rôle of a *term conjunction*, and the other being a *term disjunction*. This means that the first one has to be an *inf* and the second to be a *sup* with respect to a *unique* ordering relation $\leq$. The coincidence of both corresponding orders is ensured by two additional axioms connecting $^\circ$ and $\bullet$:

Fact 8: *In any structure with two semi-lattice operations* $^\circ$ *and* $\bullet$ *satisfying the laws of absorption*: $s^\circ(s \bullet p) = s$ *and* $s \bullet (s^\circ p) = s$, *both operations define a common ordering* $\leq$, *one of them being an inf and the other being a sup with respect to it.*

Of course, the resulting system $(=, {}^\circ, \bullet)$ coincides with the algebraic structure now named *lattice*, a structure introduced and studied for first time by Peirce (*Algebra*).

Further, let us suppose that an *extreme* element e exists in the structure $(=, {}^\circ)$ with the sole property $x^\circ e = e$ for any x. The axiom for e gives $e \leq x$ (in the sense of the ordering defined first). So e is the *least* element of the structure. The second syllogistic relation θ can be defined by $x\theta y$ *iff* $x^\circ y \neq e$. In such a case, if only elements different from e are admitted, $x\theta x$ is exactly the requirement $x \neq e$, and the analogue of the fourth Łukasiewicz's axiom, *Datisi*, follows using $e \leq m$. When the second ordering is taken, the only difference is that e appears as a *greatest* element of the structure. Let us summarize these results in the next

Fact 9: *In any semi-lattice* $(=, {}^\circ, e)$ *with an extreme element both syllogistic relations* $\leq$ *and* θ *are definable.*

Leibniz has used in *Difficultates quædam Logicæ* literally the same definitions of $\leq$ and θ calling our $\neq e$ "est Ens" (Couturat, Opuscules, 212); the requirement for any term to be non-empty has appeared as "A est Ens (ex hypothesi)" (*ibid.*, 213). Obviously, if *term negation* $(-)$ appears in the semi-lattice with minimal suitable properties, then the extreme element becomes definable by Leibniz's "non-Ens" $b^\circ(-b)$ (Couturat, Opuscules, 259). The last basis could be characterized by a triple $(=, {}^\circ, e({}^\circ, -))$ because e is represented here as a function of $^\circ$ and $-$.

When the structure contains term negation, so to say, in its full volume, the relation θ becomes definable by $\leq$: $x\theta y \leftrightarrow x \Box -y$ (as well as $\leq$ by θ : $x \leq y \leftrightarrow$ not $x\theta{-y}$). If empty and universal terms are prohibited, then the full syllogistic of negative terms can be produced from the system $(\leq, -)$ applying

Fact 10: If the negation satisfies the laws: $--x = x$, $x \leq y \leftrightarrow -y \leq -x$, $x \leq y \rightarrow x \Box -y$ (or, the last one replaced with $x \Box -x$), the traditional syllogistic enlarged by term negation can be obtained.

The last systems were axiomatized in (Wedberg, *Classes*) and (Shepherdson, *Interpretation*), respectively. Leibniz has listed the first three laws of negation, e.g., in *Generales Inquisitiones de Analysi Notionum et Veritatum* (Principles 96, 93, 91(=100)) (Couturat, Opuscules, 379–380); another exposition including the fourth formula has been given in *Fundamenta Calculi Logici* (*ibid.*, 422). When arbitrary terms are allowed, the last axiom has to be replaced with $x \leq -x \rightarrow x \leq y$ (Shepherdson, *Interpretation*).

The next combination includes term composition and term negation (besides equality) with the usual axioms sufficient to obtain a Boolean algebra (Birkhoff, *Lattice*). Then both relations $x \leq y$ and $x\theta y$ can be defined by $x^\circ y = x$ and $x^\circ(-y) \neq x$, respectively. Therefore, the system $(=, {}^\circ, -)$ is characterized by

Fact 11: *Equality, composition, and negation are sufficient for building up the full Boolean syllogistic.*

Coming back to the traditional letters and using Leibniz's notation for the composition, we can find the corresponding symmetric definitions in *Primaria Calculi Logici fundamenta*: sAp is $s = sp$, sEp is $s = s(-p)$, sOp is $s \neq sp$, sIp is $s \neq s(-p)$ (Couturat, *Opuscules*, 236).

Observing the consecutive variations in the language of syllogistic one can note a transfer from *syllogistic relations* to *term operations*, and this transfer means, in fact, a consecutive elimination of the traditional syllogistic: while only specific term relations ($\leq$ and θ, i.e., A and I) have appeared at the beginning, only term operations ($^\circ$ and $-$, i.e., composition and negation) together with term equality appear at the end. Moreover, even the *relation* = can be replaced by a *property* "$= e$" with e being the extreme element:

Fact 12: *Composition, negation and equality to the extreme element are sufficient for building up the full Boolean syllogistic.*

Indeed, if an empty term 0 is taken in the role of an extreme element and if Leibniz's notation is used for composition (now, term conjunction), then four syllogistic relations corresponding to the traditional categorical propositions can be defined in the following symmetric manner: sAp is $s(-p) = 0$, sOp is $s(-p) \neq 0$, sEp is $sp = 0$, and sIp is $sp \neq 0$. These representations have been listed, e.g., in *Generales Inquisitiones*... (Principle 151) where our $\neq 0$ is named "est res" and $= 0$ is "non est res" (Couturat, *Opuscules*, 393), as well as in *Primaria Calculi Logici fundamenta* where "est Ens" and "est non-Ens" stay for $\neq 0$ and $= 0$, respectively. It is curious that the same four equations would be proposed by G. Boole 160 years later (Boole, *Analysis*).

REFERENCES

Birkhoff, Garrett. *Lattice Theory*. 3d ed. Providence: Amer. Math. Soc., 1967.

Boole, George. *The Mathematical Analysis of Logic, Being an Essay towards a Calculus of Deductive Reasoning*. Cambridge: Macmillan, 1847.

Boolos, George, and Richard Jeffrey. *Computability and Logic*. 3d ed.Cambridge: Cambridge Univ. Press, 1989.

Bunitsky, Evgenij. "Some applications of mathematical logic to the theory of g.c.d. and l.c.m." *Vestnik Opytnoy fiziki i elementarnoy matematiki* 274 (1899), Odessa (*Russian*).

Couturat, Louis. *La logique de Leibniz d'après des documents inédits*. Paris: F. Alcan, 1901 (*reprint*, Hildesheim: Olms, 1969).

Couturat, Louis. *Opuscules et fragments inédits de Leibniz, éxtraits des manuscrits de la Bibliothèque royale de Hanovre*. Paris: F. Alcan, 1903 (*reprint*, Hildesheim: Olms, 1961).

Kleene, Stephen. *Mathematical Logic*. New York: John Wiley & Sons, 1967.

Leibniz, Gottfried W. *Sämtliche Schriften und Briefe*, Reihe 6, *Philosophische Schriften*, Hrsg. von der Leibniz-Forschungsstelle der Universität Münster, Bd. 4, 1677 – Juni 1690. Berlin: Akademie Verlag, 1999.

Łukasiewicz, Jan. *Aristotle's Syllogistic from the Standpoint of Modern Formal Logic*. 2d ed. Oxford: Clarendon Press, 1957.

Peirce, Charles. "On the algebra of logic." *Amer. J. Math.* 3 (1880): 15–57.

Shepherdson, J. "On the interpretation of Aristotelian syllogistic", *J. Symb. Logic*, 21 (1956): 137–147.

Skolem, Torlaf. "Über gewisse Satzfunktionen in der Arithmetik." *Skrifter utgitt av det Norske Videnskaps-Akademie i Oslo*, 7 (1930).

Słupecki, Jerzy. "Z badan nad sylogistika Aristotelesa", *Travaux de la Société des Science et des Lettres de Wrocław, sér.* 6, Wrocław, 1948 (*Polish*) (= "On Aristotelian syllogistic." *Studia Philosophica* 4, Poznan, 1949/50: 275–300).

Sotirov, Vladimir. "Arithmetizations of syllogistic à la Leibniz." *J. Appl. Non-Class. Logics* 9, n. 2–3 (1999): 387–405.

Sotirov, Vladimir. "The Leibniz programme: calculation in lieu of disputation." Mathematics and Education in Mathematics, 29th Spring Conf. Union Bulg. Math., Lovech, April 3–6, 2000, Proceedings, 36–45 (*Bulgarian*).

Sotirov, Vladimir. "Monadic predicate calculus with equality arithmetized à la Leibniz." *C. r. Acad. bulg. Sci.* 54, n. 1 (2001): 9–10.

Wedberg, A. "The Aristotelian theory of classes." *Ajatus*, 15 (1948): 299–314.

MARTIN TABAKOV

THE LOGIC BETWEEN TWO CENTURIES

This paper examines questions that are essential to the philosophical interpretation of modern logic. The development of logic is evaluated in the light of popular conceptions - of Kuhn - about the scientific revolutions and Lacatos - about the principles of proliferation. I analyze and generalize the philosophical considerations that forced two different revolutions in modern logic - the transition from traditional to classical logic and the transition from classical to non-classical logic and examine the mutual interweaving of logical and mathematical ideas. The relationship between the results of the Modern Logic and basic philosophical categories and conceptions are considered. The reason in both revolutions is that the development of the logical empirical sphere has gone considerably ahead of the logical theory.

I analyse critically the important moments (micro-revolutions) of the second revolutionary transition and discuss the philosophical challenges of both revolutions - Gödel's theorems and respectively the proliferation and questions about monism and pluralism of logic after the the second revolution - and offer the idea of "logical neofundamentalism". As a result of analysis, a new synthetic definition of logic is given, which encompases the relationship among logic, thinking, truth and inference and also with language and ontology. "Logical neofundamentalism" also treats the problem of the universality of logic and about structuring the area of the non-classical logics. A classification of Nonclassical Logic, consistent with the philosophical evaluation of the systems and related to a philosophical interpretation common for all logics, is proposed and critically recontextualized. The comparative analysis contributes to understanding the place and the importance of each separate logical system - in some sense non-classical logic's (and also classical logic) "are not what they are". The transition from classical to non-classical logic is also related to the transition from classical to non-classical science.

1. THE TWO REVOLUTIONS IN CONTEMPORARY LOGIC

Two different revolutions in modern logic.

The great development in contemporary logic has been recognized even by those who do not know well the new logical studies research and who, maybe for that same reason, have a more skeptical attitude to some of them. In evaluating this development from the historical and methodological point of view, the word

Dimitri Ginev (ed.), Bulgarian Studies in the Philosophy of Science, 95-126.

'revolution' suggests itself as the most exact description: in a sphere of knowledge that used to be deemed almost completed, whose basic postulates (dogmas) have been viewed as doubtless and have almost been canonized, totally new conceptions emerge! 'Revolution' corresponds to the scale of change and reevaluation of values in modern logic, comparable to important moments of the development in other fields: Newton's theory, the periodic table of the elements, quantum mechanics, the transition from the Newtonian physics to the Einstein's theory of relativity, non-Euclidean geometry [1].

Such an evaluation is found in authors having a mostly mathematical logical orientation as well as in authors having a philosophical (epistemological) orientation. However, most authors reach it and stay there, without analysing further what it is that the revolution precisely consists of, what stages has it passed through, and whether or not this is a case of two different revolutions. The development of contemporary logic is evaluated on the large scale, with no specifying or going onto the details of its separate stages. In my view, there is a revolution in contemporary logic, but it consists of two distinct parts: a transition from traditional logic to the contemporary classical logic, and a transition from classical to non-classical logic. The difference between them is so outstanding that, from the philosophical and methodological point of view, it would be more precise to speak of two revolutions, both leading to a replacement of one paradigm by another.

An essential cause for these revolutions is that logical empirical development has gone considerably ahead of logical theory. It has been debated many times whether logic has an empirical base or is a purely deductive structure. Quite a number of authors argue that logic is a science without such a base. On this question, so important for the philosophy and methodology of logic, my position is that logic does have 'empirics' and this is basically the language and methods of reasoning of scientific theories. Deduction is of course applied most consistently in mathematics; therefore the language and methods of reasoning of mathematics are empirical material for logic. However, it would be groundless to restrict the logical empirics to reasoning in mathematics alone; the language and methods of reasoning in many non-mathematical theories are empirical phenomena too. Logic deals with the transfer of truth from the premises to the thesis in an objective and non-empirical way but this 'non-empirical transfer' has its own empirics. It has often been argued that the concepts and ideas of logic are based on experience. According to Bacon, logical forms and human thinking coincide with the world for the same reason by which our lungs are suitable to the atmosphere of the Earth. I believe, however, that there are intermediate stages even here. There is something of the kind in logic as well but the number of intermediate stages is greater. If the empirical base of logic consists of the reasoning and justifiability (provability) in scientific theories, one should not forget that in their part they are empirical results as well. It would seem that the Empirical of logic is logic itself but this is not the dangerous, paradox-generating self-reflection: the demarcation is clear into reasoning and the theory that studies them. In the 18th and 19th centuries, the very rich and most developed empirical material for logic was provided by mathematics. This is why mathematical reasoning was taken as the basis for the construction of the first logical systems. In mathematics, the Empirical was rich enough to be generalised and explicated into a logical system.

The 'empirical-theoretical' relation has been connected to the concepts of 'verification' and 'falsification'. In logic, the Empirical has a more specific character; one cannot say that it provides a direct 'verification' and 'falsification' to the theory. There is here another theory as an intermediary. However, in taking into account this intermediary, one can speak again of 'verification' and 'falsification'. For the logical theory called 'classical logic', the mediating theory is mathematics, and its empirical base - the logic of mathematical reasoning - verifies classical logic. However, in classical logic, the paradoxes of material implications obtain, while for the bulk of scientific theories such a type of logical reasoning is unacceptable, and this is why, in my view, we can think that the logic of such a type of theories (i.e., the empirics) 'falsifies' classical logic (or, at least, puts its universality in doubt).

The lag of the science of logic behind the logical empirics, and especially behind the data provided by mathematical reasoning, started long ago. As early as in Euclid's works, many more logical forms and constructions are used than those studied by Aristotle. They fall outside the scope of Aristotelian syllogistic and cannot be described in its language. For example, in formulations and demonstrations, multiple predicates are used such as 'between' etc. Euclid largely uses the classical propositional logic's operations - disjunction, conjunction, negation. Subsequently, logical theory has for a long time been lagging behind the mathematical logical empirics. Especially striking was the lag in the 17-19th centuries when mathematics underwent a rapid development. The logical theories then available did not provide the link between logic and mathematics. The paradigm of traditional logic had exhausted its capacities long before the first revolution finally happened. It was not until the works of Boole that logic started to catch up with mathematics, and with the works of Frege, Russell and Hilbert, it overtook it and currently has even come slightly ahead in some details: in contemporary logic, logical constructions are used that surpass the logic of contemporary mathematics.[2] The works of George Boole, De Morgan, Gottlob Frege and David Hilbert are the result of the generalization of the logical empirics of mathematics. They eliminate the mentioned lag. This is an example of how logical theory generalizes logical empirics. For that matter, since mathematics is related to (and is the result of) extensional theoretical structures that lie outside it, the mentioned results are the generalization of an empirical base encompassing a larger area.

The revolution - *from traditional to classical logic* - consists more of the change of solutions, instruments and methods. The system of classical logic uses a wider range of mathematical instruments, the axiomatic method, it allows for working with multiple-place predicates, it provides the possibility to unite spheres that have theretofore been conceived, in logical theories, in isolation, each for its own sake. For me, it is precisely this possibility that represents the most significant result of this revolutionary transition. The old tradition, after existing for a considerable time without changes, becomes replaced by a new tradition having its own language and based on a new conception of expression and study of logic. The new paradigm has almost completely replaced the old one.

The question is more peculiar as to how far the new paradigm is commensurable with the old one. The system of classical logic does not deny traditional logic. It generalizes and essentially extends its capacities. This is why I believe that we cannot speak here of a complete incommensurability. Some of the orthodox followers of Kuhn would conclude that, therefore, there could be no talk of revolution here. However, in my view, a comparable absence of incommensurability can be also observed in some changes that have been recognized as revolutions in science, especially in the cases when one goes from some separate theses and works to a complete large-scale theory. Newton's theory is of such a type. In a greater degree, incommensurability arises in the second revolution: the transition from classical to non-classical theory. Here, one talks much more often of incommensurability and most scholars seem to accept that overtly or implicitly. My opinion is slightly more specific. There is doubtless incommensurability but it could be sublimated into a suitable conception of logical neo-fundamentalism.

The *transition from classical to non-classical logic* is another revolution, and a much more important one, since it affects issues of a fundamental nature. This is not a question of introducing qualitatively new methods; there are qualitatively new methods in the first revolution while here, there only is a certain extension of these methods. The question here is of qualitatively new conceptions, ideas that radically change our understanding of logic. In the first revolution, the introduction of new, mathematical methods makes theory clearer and more all-encompassing, logical theory comes, in its style and method, closer to mathematics - the field where the empirics has had the greatest need for a new logical theory. Although coming rather late in time (many of the causes have also a socially psychological nature), this revolution is comparable to the works of Newton and the creation of classical physics, and also to the creation of the periodic table of the elements. (The analogy here comes naturally - the transition from classical to non-classical logic can be compared to the theory of relativity and quantum mechanics). In its spirit, it is a fundamentalist revolution giving new foundations to logic. The second revolution, on the other hand - the transition from classical to non-classical logic - is strongly antifundamentalist in its nature; it puts under doubt this same basis that the first revolution has just propounded. It too has been provoked by the discrepancy between theory and empirics. However, the empirics in question here is extra-mathematical: it is the logic of scientific research. Actually, the development of contemporary logic seems to me natural and consistent here: the initial breakthrough is done in the area where logic is used most extensively, most systematically and most coherently, in the use of its means of expression. Becoming aware of the efficiency and productivity of contemporary methods, contemporary logic continues to develop but this time it does it by going deeper and wider: mathematical methods allow it to explicate also essential logical moments of the extra-mathematical logical empirics.

To sum up, I need to say that while the first revolution - the transition from traditional logic to the system of classical logic - is a revolution of methods, the second revolution - the transition from classical to non-classical logic - is a revolution of ideas and postulations.

An important reason for why the two revolutions have been viewed as one whole is that they are very near in time to one another and in fact happened almost simultaneously. Actually, it seems to me that Boole's idea to introduce algebraic methods into logic, reducing logic to a simple algebra, could well emerge considerably earlier. The works of Frege, however, are historically related to the development of science. They come in response to the need for logical analysis of the foundations of mathematics, and thus are the natural development of the creation of the set theory and its establishment as the basis of mathematics. However, non-classical logic comes to light very soon after them. The system of classical predicate logic had yet not been completely created when Lukasiewicz proposed his system of three-valued logic, and the system of Lewis emerges at that same time. It is only natural to perceive them as a continuation of the works of Frege and Hiblert. The methods are similar: formal and mathematical. (To try to make free analogies, we could imagine that Newton's theory historically came much later and was created only by the late 19th. century.) This has been another obstacle to their methodological demarcation. It is only too easy to view them as one whole or as one natural process, as two stages of the same development. Of course, if there should have been a transition from classical to non-classical logic and if the real construction of systems, alternative to classical logic, was to happen, then classical logic itself must have been completed, at least to some extent. Heyting's system of Intuitionistic logic was created naturally after the creation of several axiomatic systems of propositional logic. However, it was not at all absolutely necessary for the classical logic to be constructed up to its total completion before its competing alternatives came into being, moreover the idea of non-classical logic come much earlier.

Another general reason to view the two revolutions as one whole is that the second is a continuation of the first: besides being temporally near to one another, they also have similar causes that generated them. Both are the result of the lag of theory behind the empirics. The difference is that in the one case the empirics is mathematical while in the other it is non-mathematical.

Logical paradoxes are a third general reason: the paradoxes in set theory, semantic paradoxes, and the paradoxes of material implication. However, in classical logic they exert their influence at a later, intermediate stage. The paradoxes in set theory are a catalyzer; they affect principles and postulations that are basic to the theory. The idea to form a set of sets is essential to Cantor's theory. In his approach, the set is an object like any other object and one can treat and manipulate it in the same way as other mathematical objects. Then, it is also natural to use the same logic. However, finite and infinite sets are too different, they have qualitatively different properties and, since they are so different, it is debatable whether the same logic may be used. While in the first revolution it was Russell's paradox that gave the greatest impetus, semantic paradoxes and especially the paradox of material implication have played a greater role in the second, demonstrating the incapacity of classical logic to express logical inference adequately. In part, the synthesis of the two types of rationality cohabiting in contemporary logic - the philosophical and the mathematical one - is the basis of revolutions in contemporary logic. This is especially valid for non-classical logics. The creation of many-valued, modal, intuitionistic logics is related to philosophical postulations. Unfortunately, this synthesis is sometimes partial, inconsistent and incomplete.

There is another synthesis in the revolutions. The Stoics used to divide propositions into simple and complex and to study how the truth in the complex ones depends on the truth of the simple ones from which they are composed and on the form and manner of their combination. Aristotle restricted himself to the analysis of some elementary subject-predicate relations. Propositional logic views propositions as indivisible, “structureless”, with the only property to be a truth or a falsity. In the system of classical predicate logic, a successful synthesis has been achieved of the two approaches to logic, which have begun their development in antiquity: Aristotle's subject-predicate approach, studying inferences based on the subject-predicate structure, and the Megarians and Stoics approach, studying inferences based on propositional relations (well realized in the system of classical propositional logic).

The first revolution detached logic from ontology. The second provided a possibility to restore this connection. The ideological basis of the first is Leibniz's ideas to present demonstration as a calculation similar to the mathematical one, and to introduce more systematically mathematical methods into logic. Retrospectively, the gist of the idea is the view that it is precisely the methods of this type that would be efficient to the logician. The ideological basis to the second revolution is provided as early as in Aristotle's modal syllogistic, the works of Ramos, and - unexpectedly to some - the spirit of the works of Hegel.

Whenever speaking of revolutions, the two views emerge immediately: the revolutionary and the evolutionary one, namely, 'revolutions are gradually and continuously prepared evolutions', and 'however well prepared, revolution is revolution'. We face this question both when speaking of social development and when discussing issues of the philosophy of science. My aim here is not to consider this question in detail, but I still think that here too, a synthesis is possible between the two views: revolution is the result of evolution. In the revolution, the evolutionary potential becomes realized quickly, in a relatively narrow period, leading to a considerable change in a relatively short time. For a number of specialists in social philosophy, the cause of a revolution is the retarding or retention of the natural evolutionary development of society, and revolutions are the symptom of an unsound state of society, of essential errors in its development. If it had developed in the natural evolutionary way, there would be no need for a revolution. I might accept this position where society is concerned but I do not think that it could be mechanically applied for the revolutions of thought. I would not be able to perceive the emergence of a grandiose idea changing many basic postulations as a symptom for an 'unsound state' of the given field or for 'essential errors' in its development. However, I see these words as appropriate when such an idea has come too late and thus the respective field has lagged behind other fields related to it! The revolutions in contemporary logic have been conditioned by the considerable advances in the development of logic behind mathematics and the other sciences related to it. At a certain moment, it has become necessary for it to catch up within a very short term. There is a link between the development of ideas in the different sciences. Many discoveries in physics have been related to results of mathematics, and vice versa. Achievements in chemistry have been related to results in physics. This internal correlation in the development of the separate sciences exists especially in fields that penetrate in a specific way into many areas of knowledge. A clear and often used example for that is mathematics: one can discover mathematical regularities in many areas of knowledge, mathematical models have a wide application. Maybe less often mentioned, logic is such a field with a considerable wide application. This is a reason for the lag to be felt as especially painful.

The revolutions in logic raise questions in a spirit close to that of the revolutions in physics: what would be justly deemed as a revolution; is there one or are there two revolutions. Philosophers of physics ask themselves 'whether one can view as revolutions the discovery of X-rays or Mёssbauer's effect...'. A question that seems similar to me is whether Gödel's restrictive theorems are a revolution. If the emergence of classical dynamics (Galileo and Newton) is a revolution, then the emergence of the system of classical logic is, for me, an achievement in logic similar in spirit, and this is why it too should be called a revolution. The question whether it is a matter of one or two revolutions is not only raised in logic. A similar question has also been asked about the development of physics. The relativist and the quantum revolutions: are those two separate or one common relativist-quantum revolution? Is the theory of relativity a continuation of the revolution in electrodynamics, or it is separate as a subsequent revolution? Is this one, or two revolutions, of the special and of the general theory? (Polikarov, *Revolutions*)

Another analogous point: the theory of relativity does not deny or replace classical physics. Analogously, non-classical logic does not deny or replace classical logic. Insofar as the philosophy of physics accepts the thesis that the theory of relativity is more suitable to some theories related to certain entities and ontology, similarly the philosophy of logic could accept the thesis that for some theories related to certain entities and ontology, some kind of non-classical logic is more suitable.

In the development of logic, another moment is observed that is characteristic of revolutions: a rethinking of the universality, of the scope of certain basic principles and postulations. There are several and not just one such moment, and in each of them the sphere of validity of some basic principle or postulation undergoes a change. For some, this is a sufficient reason to talk of a revolutionary transition, but while the word 'revolution' is nevertheless too strong, I deem as most appropriate and I am using the concept of 'micro-revolution' for any such rethinking of a basic principle of classical logic. The transition from classical to non-classical logic consists of many different stages or micro-revolutions. Micro-revolutions do not happen in a single turn, they too have a more complex nature; they consist of different steps that are sometimes so small they resamble evolution. Micro-revolutions are not always historically consequent; they do not strictly follow one another. In the historical development of logic, some micro-revolutions have taken place parallel to others and some have even come before others. The ideas of Vassiliev - the forerunner of intensional and paraconsistent logic's (Arruda) - came before the emergence of many-valued and modal logics. The idea of extending classical logical operators by the modal ones seems to me more natural and acceptable from the classical point of view and much more important and desirable from the philosophical point of view. However, there is awareness of it only after Gödel's axiomatisation of the systems of Lewis - later than the creation of Intuitionistic and many-valued systems. Relevant logic (see Anderson and Belnap, *Entailment*), which is much more closely related to logical inference, has emerged only after it has become clear that modal logic is not suitable for that purpose.

In classical logic, several stages of development can also be indicated. The first is the creation of Boolean algebra, which gave a precise and complete form to the ideas of propositional logic that had been rooted in the works of the Megarians-Stoics thinkers but not well developed for quite a long time. Why that happened, is easy to construct hypotheses about but hard to justify. In my view, the idea could well have been developed as early as in Ancient Greece. The Stoics divided propositions into simple and complex and studied how the truth in the complex ones depends on the truth of the simple ones from which they are composed and on the form and manner of their combination. However, this promising approach has remained in the shadow of Aristotle's syllogistic and historians do not explain why the works of the Megarians-Stoics school, which became the forerunner of propositional logic, have undeservedly stayed at the side of the mainstream. A role was of course played here by Aristotle's authority, his style and systematics, and maybe also socio-psychological factors, upon which we may only conjecture. The mathematical realization of these ideas also came too late. As a mathematical structure, the Boolean algebra could well have emerged at least a century earlier: symbolic notations in algebra were used successfully in the 16th and 17th centuries. To compare, the theory of groups, generalizing a much later developed and relatively more complex mathematical empirical basis, emerges with the works of Niels Abel (1824-27) and Evariste Galois (1829-30) before Boole's works (1847, 1854). The algebrization of logic is a considerable micro-revolution in the development of logic. A following stage, organically related to the idea of the previous one, is the construction of propositional logic as a logical system. Finally, the stage concluding this revolution is the construction of predicate logic as a logical system (Frege, Hilbert).

Philosophical challenges of the first revolution.

The famous Gödel's incompleteness theorems are among the most important methodological results as a consequence of the first revolution. A thesis popular among the mathematical community is that these theorems are in themselves a revolution. However, their condition is the creation of the system of classical logic. This is an interesting point: a new theory is built and it allows one to obtain an important result. If we believe that there are reasons to call this result a revolution, when did it start, then - with that result or with the creation of the new theory upon which it was conditional? I am more inclined to call the creation of classical logic a revolution. There are quite a number of important results in physics that were conditional upon Newton's theory but they are not called revolutions. Besides that, although historically Gödel's theorems were not expected, something of the kind could be expected retrospectively, since Hilbert's program allows for the situation of self-reflection similar to some semantic paradoxes. The consistency of arithmetic is proved through arithmetization - the system contains arithmetic and must be proved through it, i.e. it contains both its own syntaxes and semantics. Gödel's theorem concerns an arithmetical statement and the truth of this statement does not depend on self-referents, nevertheless its interpretation is of tremendous interest. Gödel's theorem allows one to conclude that formal arithmetic is not axiomatizable. However, there is another very interesting logico-philosophical result: the possibility of a statement to exist such that it is unprovable in the object-theory and at the same time its truth is provable in metatheory. It seems that in the real history Gödel's theorem was absolutely unexpected, and even in retrospect, it is now difficult to point out logico-philosophical reasons for the above possibility. In my opinion though it was still conceivable at that time, but unfortunately the philosophers and logicians failed to notice it. In the terms of recursive arithmetic Hilbert's program can be formulated as follows: build a decidable, complete, consistent (even categorical) logical system such that the class of derivable statements coincides with the class of mathematical statements that is intuitively true: prove this equivalence by means of an arithmetization of the theory in the framework of recursive arithmetic! Now, if the system is so strong as to admit the construction of (a large part of) classical mathematics, then particularly, it will contain classical arithmetic. Hilbert's requirements reduce to a proof with the means of recursive arithmetic, by arithmetization.

So, on the one hand the system will contain arithmetic, and on the other it will be expressible in it, in other words the system will include its own syntax, as well as its own semantics. Now it is clear that Hilbert's program contains a possibility of a situation similar to some well-known semantic antinomies, the situation of self-reference.

Here we put forward an argument that shows that Gödel's theorem is not that unexpected. We first consider the problem of the subject of logic and the notion of logical truth. Logic is interested in such kinds of reasoning that transform true premises into true conclusions. A logically true (valid) statement is such only on the basis of its logical form and the meaning of the logical terms that appear in it. One of the aims of logic is the study of truth with respect to the chosen meaning of the logical terms - a change in this meaning changes the class of valid statements. At the same time logical truth does not depend on the meaning of non-logical terms. Logic investigates which statements are valid, i.e., true irrespective of the non-logical terms and their interpretations. So the logic valid in an object domain is characterized by the class of logical truths therein. Conversely, logical truths are universal in the object domain and true in every world therein, hence they cannot be distinguished by, say, physical or whatever non-logical means. (See also Tabakov, *Goedel*)

As far as we are concerned, with first order logic there is completeness theorem, which shows that universal truths can be defined syntactically. The situation with formal arithmetic is quite different. Here we try to obtain not the statement valid under all interpretations but the statements true under a particular interpretation (the so-called standard interpretation). It is the class of these statements that is not expressible in the system. In this way we reach a conclusion: systems defining statements valid in an object domain can be built, but often it is not possible to built systems for statements true in a particular interpretation. We feel that this is not unexpected, since logic studies only those truths that are universal. It follows that the results of Gödel and Tarski, far from being unexpected, are in fact quite natural indicators of the inability of logic to treat concrete truth. This is in perfect accord with the conception of logic that we hold.

Gödel's theorem can be viewed as a consequence of a very general thesis. Finitary methods eliminate the risk of errors. Gödel's theorem shows that finitary methods miss some of the truths. If we stick only to absolutely reliable methods we have no chance of reaching certain truths. Knowledge acquisition is inseparable from the risk of errors and any attempt to eliminate this risk necessarily leads to limitations upon our cognitive resources. The cognitive process inevitably passes through stages that are not certain.

Contemporary logic is closely connected to the development of the philosophy of mathematics. The crises in the foundations of mathematics are among the important factors for the development of contemporary logic, more especially for the revolutions discussed here. The creation of the system of classical logic was provoked by the Third Crisis in the foundations of mathematics (the late 19th and early 20th centuries) related to the paradoxes of the set theory, and just at the moment when it was expected to become a stable foundation to the construction of mathematics. That same crisis also provoked intuitionism and, with that, intuitionistic logic. We face here an interesting point, maybe not very standard to the philosophy of science. A standard postulation would be: a crisis in a certain field leads to a revolution in that same field. In the case of the revolution in logic, things are somewhat different: a crisis in mathematics leads to a revolution in another, albeit very close and related field - logic! An attempt to standardize the situation would be to accept that in fact it was not a crisis in mathematics but in the logical foundations of mathematics. However, set theory is nevertheless a 'purely' mathematical discipline. Almost everywhere philosophers of mathematics talk of a 'third crisis in the development of mathematics'. They also talk of an earlier such crisis caused by the discovery of irrational numbers, and view it as being as important as the third one. This means that they are of the same type: crises in the development of mathematics, crises in the foundations of mathematics. However, in the case of irrational numbers I see no direct relation to the 'logical foundations' of mathematics. On the other hand, the development of logic has led to the rethinking of many questions of the philosophy of mathematics. The first revolution, for instance, imposed upon working mathematicians a structuralist conception of mathematics: 'With its development, mathematics becomes more and more abstract and, from a "science of quantitative relations and spatial forms", it has turned into a science studying abstract mathematical structures, mathematical models of theoretical systems.'

The micro-revolutions in the second revolution.

With the construction of the system of classical logic, logical theory corresponds to and satisfies the everyday needs of mathematics. However, at that level, logic as a theory still lags behind and does not correspond to the level of most related sciences (in whose reasoning logic should find its empirics). To make it able to become closer to them, it needs some of the subsequent micro-revolutions. What is common to the other micro-revolutions, is that they are a part of the second large revolution: the transition from classical to non-classical logic. It seems strange but despite its obvious linking points with philosophical problems, it did not gain understanding everywhere among philosophical circles. The development of modern Logic is estimated as a revolution, consisting of a series of successive separate stages (micro-revolutions). In my opinion the important moments of revolutionary transition (micro-revolutions) are emphasized as follows.

a) Abandonment of the principle of the excluded middle (Tertium Non Datur) - Intuitionistic logic. Intuitionistic logic encompasses the principles of logical reasoning, which were used by L. E. J. Brouwer in developing his intuitionistic mathematics. (Moschovakis)

From the Non-Classical Logics, Intuitionistic Logic is among the most established logical systems. Most of the Logicians-philosophers do not dispute its status as a Logic, and those who do, dispute such status of any Non-Classical Logic. However the question of the philosophical and the methodological rationalization of Intuitionistic Logic is still open. To what extent does Intuitionistic Logic correspond to the philosophy of intuitionism and are the proposed semantics adequate for this philosophical standpoint? Is it Logic of and for mathematics, or does it refer to many other theories and fields? Does it explicate the constructive thinking only in mathematics, or does it explicate the constructive thinking in a considerably wider philosophical sense? Is it the only possible constructive Logical conception, or there are other worthy alternatives?

I emphasize Brouwer's idea of the connection of Logic with the object of study, to which it is applied - Logic could be modified by the change of the object of study. It is limited only to the Logic of and for mathematics. The idea of the absolute universality of Logic is rejected, which poses the problem of the pluralism of the Logic. It is possible to apply different logic in theories studying different entities. There are cases when some entities necessarily imply the use of non-classical logic for the theories that study them. The basic question is how to set apart such non-standard entities from the remainder, in which Classical logic could be comfortably applied. If obstacles are met in a certain area, its logic must be examined. When the work with a logic not only leads to a great complicating of the exposition, but also is an obstacle to the completeness and the adequacy of the cognition, this logic may not be adequate to the ontology. Something similar applies with Intuitionistic Logic - the crisis in the foundations of mathematics stimulates the search for its solution in change of the logic. Intuitionistic Logic is the only Non-Classical Logic, on the basis of which set theory has been developed seriously and profoundly, and on its basis, other branches of mathematics, too. We may assume that Intuitionistic Logic corresponds to the ontology of mathematical objects. It is acceptable to speak about the ontology of mathematical objects, and about an ontology of some abstract objects in general. If we assume that (only) one logic corresponds to a definite ontology, it follows that two different ontologies exist - ontology of the objects of classical mathematics and ontology of the objects of intuitionistic mathematics. Intuitionistic Logic is not the strongest refusal of the law of excluded middle. A more categorical rejection is given by Many-valued Logic. However, both types of Logic should not be considered at one level unrelated with their philosophical interpretation. Intuitionistic Logic serves as the basis for a mathematical theory; this is its most important interpretation. Many-valued Logic has a completely different philosophical substantiation, related to epistemological grounds. It is possible, in a certain epistemological situation, that the law of excluded middle may be invalid in a much stronger sense, than it is in Intuitionistic Logic. A Logic that rejects this law in a much stronger sense than Intuitionistic Logic could not do a serious job in mathematics. It is interesting to construct an alternative Logic, based on epistemological considerations of "constructiveness" of the Truth. Not the mathematical truth, but the truth as a philosophical category! It is possible that there may appear a situation in which the thesis "it is impossible not to exist, consequently it exists", may not be satisfactory. For such an alternative Logic mathematics would be a particular and perhaps insignificant case. The purpose of such a Logic will be to construct and analyze a Logic, which not only determines the truth, but also effectively indicates the way to it. This may be Logic related with situations from the social sciences, with management, decision making, the legal system, related to situations in which the important is not "what", but "how" (Tabakov, *Logic*). Constructive Logic allows for a more detailed classification of propositions, which could be related to some epistemological situations. The idea of the differentiation of true propositions - common truths and constructive truths - is similar to the Modal Logic idea of the differentiation of truths - common and necessary truths. Classical physics works well with Classical Logic, but in the special theory of relativity the criterion of simultaneity of events is more constructive. Hence, a constructive Logic will be needed in the analysis of objects, related to such simultaneity of the events.

b) Abandonment of the principle of "2 (and only 2) values", one of the basic and most important principles of classical logic - Many-valued logic does not restrict the number of truth values to only two: they allow for a larger set of truth degrees (e.g. Gottwald; Rescher; Rosser and Turquette). This enormous leap alone would suffice to justify the word 'revolution'! There is a large group of logical systems in which the principle of two values does not obtain in one sense or another. In fact, this refers to almost all non-classical logics. However, for some non-classical logics multiple values are a basic aim while in other logics multiple values come out as the consequence of other, different postulations such as: distinction between 'ordinary' and 'necessary' truths; avoiding the paradoxes of material implication; requirement of a constructive nature of mathematical demonstration. This is why a classification has established itself in the literature, according to which 'many-valued logic' includes those logical systems whose basic and most characteristic trait is the abandonment of the principle of two values. Those logical systems that are essentially many-valued but whose basic characteristic is something else become separated into another class of logics in accordance with their main characterizing property: modal logic, intuitionistic logic etc. The main idea is that two-valued logic cannot express the diversity of logical situations, e.g., in propositions of modality or probability. Even Aristotle has noted that, for future events, some laws such as the excluded middle are not valid. Classical logic imposes on us to restrict ourselves solely to propositions whose truth value is categorically known: an idealization seldom met outside mathematics, as in most Epistemological situations the only two evaluations of 'truth' and 'falsity' are too coarse. It can be said of many statements that they are true only to a certain extent, but the indefiniteness of their truth does not exempt us from the need to perform logical operations with them.

Most of the first many-valued logics (the systems of Lukasiewicz, Heyting, Bochvar) have three values. Depending of the Epistemological situation to which the logical system is related, admitting a third value having a truth status different from 'true' and 'false' allows for a rich palette of interpretations: "neutral", "indefinite", "possible", "unknown", "predefined", "probable", "antinomic", "meaningless", "overloaded" etc. These interpretations of the middle value are manifold and, what is more important, quite different from one another. In my view, some many-valued logics are, rather, mathematical models! I accept the possibility for a system to have many and different interpretations but, in this case, the great number of diverse interpretations of the third value do not increase the philosophical and methodological importance of the system; they also have a different semantic and general philosophical interpretation and this is why it would be logical for them to find expression in different systems, taking into account the specific nature of every separate interpretation. It is a matter of philosophy to answer the question of which interpretation of the third value best corresponds to, say, the logic of Lukasiewicz.

The system of Lukasiewicz is, rather, a mathematical model of a logic having a third value. It is an explication of the general idea of a logic with a third value but it cannot be accepted as a good explication of some of the mentioned specific interpretations. Its main methodological importance is the implementation of the revolutionary transition, the rejection of the principle of two values. It is more of a starting point, a basis and direction for future research, without having too great capacity as it stands. In (Vakarelov, *Notes*, p. 111-125), an additional interpretation to the system of Lukasiewicz points out that it is the explication of a system, including a classical logic with a strong negation. Philosophically viewed, this interpretation is more important.

The skeptical evaluations of many-valued logic are not in fact directed against the idea of many values but against its implementations, which is too schematic in some systems. If we analyze the possible interpretations of the third value, we will see that in most of them it is subject to discussion whether or not the logical system that would correspond to them should have precisely three values. For some interpretations, another type of many-valued logic would seem to me more appropriate. Thus, as a logic for databases and computer systems, a logic of four values is more suitable (Tabakov, *Logics*). Just one additional value provides too few opportunities. This kind of analysis can lead to the conclusion that for a given interpretation there is no adequate logical system, which could lead to the construction of new systems that would take into account the specific nature of the interpretation. Epistemological interpretations of modal logic lead to the idea of many-valued logic, but a many-valued logic with a special interpretation of truth values related to their modal interpretation. In most of the possible interpretations of the third value, it is open to discussion whether it is the three-valued logic that best corresponds to them. For some logicians, another type of many-valued logic seems more adequate. I think that the abandonment of the principle of two values does not end with at three values: allowing one intermediate value leads to the requirement for others; most interpretations of the third value allow for different degrees, e.g., 'more indefinite'. The interpretations of the third value have a modal importance, and modal operations also allow for degrees: 'very necessary', 'more necessary', 'less necessary' etc. In my view, from the philosophical-methodological perspective, the main point is not to have multiple values with truth values defined beforehand, or multiple values generating modality, but multiple values generated by a modality extracted from the need for a logic considering a modal graduation of truth. Many-valued logical systems having fixed matrices of truth values, although they too have specific interpretations, would be classified by me as mathematical models. From the philosophical methodological point of view, probability and topological logics are much more interesting than the systems of Post. They are also closer to my interpretation, according to which the rejection of the principle of two values leads in many cases also to the rejection of the principle of three values (which is another scheme representing too strong an idealization and does not lead to systems adequate to the philosophical intuition of multiple values). For some of the interpretations of the third value, the rejection of the principle of two values also leads to the rejection of the principle of a finite number of values, which leads to an infinite-valued logic. This is another argument against the artificiality of the many-valued logic of Post.

Many-valued logics are an instrument for a more suitable and precise study of epistemological situations by logical means; they are related to the idea of distinction and graduation of the truth of propositions. Admitting the possibility for intermediate values also imposes a rethinking of views on the classical concepts of 'truth' and 'falsity'.

c) Extension of classical logical terms. Modal logic extends the classical one by operators like 'possible', 'necessary', 'contingent' etc.; the deontic logic by 'obligatory', 'permitted'; the temporal logic by such as 'always will be', 'always', 'still', 'already'; logic with descriptors by 'that A which...', 'an A which...'; axiological logics by 'good', 'bad' etc. At the semantic level, this can be interpreted as an abandonment of having just one type of truth! The requirement to consider truths of different types is another breach of the principle of two values. Besides 'truth' or 'falsity', propositions must be at least 'necessary truth' and hence also 'necessary falsity'.

d) The transition from extensional to intensional logic (Van Benthem). The extensionality of classical logic also determines the limits of its efficient applicability: theories where extensional definiteness is appropriate. However, most scientific theories involve an intensional approach. For a long time, logic has been related to a mandatory extensional treatment of objects, an abstraction from content and consideration of form alone. The apparatus of contemporary logic has surpassed these restrictive limits. It is mighty enough to encompass intensional contexts as well. Methods and mechanisms created within the framework of an extensional tradition can be also applied outside it and express moments of content as well. Intensional logics are an alternative both to classical logic and to non-classical extensional logics. Intensional logics are a realization of the wish, repeatedly expressed by different philosophers, for the creation of a 'logic of content'.

e) Abandonment of the universality of the law of non-contradiction: an important albeit most debatable step affecting a very basic law of the classical logic. The development of paraconsistent logic was initiated in order to challenge the logical principle that anything follows from contradictory premises (Arruda, Priest and Tanaka). There are many different paraconsistent logics -Non-Adjunctive Systems, Non-Truth-Functional Logics, Many-Valued Systems (Asenjo, *Calculus*; Asenjo, Tamburino, *Logic*) and so on. However, the problem concerning the philosophical interpretation of paraconsistent logic has not yet been solved. Jaskovski (Jaskovski, *Propositional*, p. 173) has pointed out the following constraints on an inconsistent, but not-trivial logic:

> "1.When applied to contradictory systems would not always entail their over-completeness.

2. Would be rich enough to enable practical inference.
3. Would have an intuitive justification."

Historically, paraconsistent logics were often used for the construction of set theories with the aim to avoid Russell's paradox. Thus, by analogy with intuitionistic logic, paraconsistent logic is regarded as an alternative to classical logic. Some authors (Da Costa, *Theory*) actually replace 2 and 3 with:

2a. The system should easily be extended to a predicate calculus.

2b. 2a should make it possible to construct a mathematical set theory).

3a. There should be formal semantics for the system.

According to us this view is not well grounded. In our opinion 2a is too strong a requirement. In a number of non-classical logics - modal and relevant ones. Interesting results are produced even at the level of propositional calculus. Moreover, modal logics with quantifiers are somewhat doubtful - grafting quantification theory on modalities raises a host of difficulties; 2b is a consequence of the view that paraconsistent logics are an alternative to classical logic; 3a also need not be considered as absolute. For a short period, the relevant logics (Anderson and Belnap, *Entailment*) had no formal semantics, but this was no obstacle to their intuitive appeal.

That is why I am of the opinion that the requirements in questions are not at all necessary for paraconsistent logic. They are mainly the result of attempts to imitate classical logic aiming to build mathematics on such a basis. However, attempts to get rid of Russell's paradox by way of logic with a weaker negation are doomed to failure. Curry's paradox shows that a construction, similar to Russell's paradox is also available in a logic without a negation operator - intuitionistic implicative logic. Russell's paradox is therefore not connected with negation, but rather with the abstraction principle. (Tabakov, *Logics*)

The difficulties in building paraconsistent set theory are not a matter of chance. Although mathematics is said to be an ontology-free science, some deeper mathematical ontologies still exist. In the same way as the ontology of the micro-world determines the logic of the sciences that deal with it, so does the ontology of mathematical objects in respect to the related theories. Intuitionistic logic turns out to be compatible with mathematical ontology, but not so with paraconsistent logic.

The importance of paraconsistent logic for philosophy is greater than for mathematics. The aim of paraconsistent logic is to study the logic of a theory, which tolerates contradictions. The necessity of such a logic is quite different from the necessity to construct mathematics to deal with it. The construction of systems, tolerant to contradictions, can be associated with the explication of real epistemological situations. The history of science could furnish instances of inconsistent theories doing their job. For some time such theories do not fall out of use, though scientists use then with caution. Many contemporary mathematicians who are not so easily disconcerted with the paradoxes use naïve set theory in their current work. In this perspective the role of paraconsistent logic could be viewed as methodological in the following sense. Paraconsistency does not open a way out of the paradoxes, as it is not the likely candidate for an underlying logic of mathematics. It rather explicates in logical terms how a mathematics (or other theory) with paradoxes does its job. If a contradiction were actually found in arithmetic - that will only prove that an arithmetic with a contradiction in it could render very good service. (Wittgenstein; see also Tabakov, *Philosophical*)

There is a philosophical outlook, according to which for some fragments of reality there can be no consistent world-picture or account. Any such account would be incomplete. This outlook is connected with so-called "Hegel's thesis" that there are "true contradictions". According to it consistency is a sufficient but not necessary condition for the existence of abstract objects, and concerning the existence of concrete objects is neither necessary nor sufficient. The "real antinomy" could not be eliminated in any normal way that would come down to a replacement of subjective elements by objective ones. Real antinomies are not fallacies but are "peculiar objective truths". This makes it necessary to produce a logic that is adequate for the purpose of studying contradictory, inconsistent entities. By the way, it can be argued with great certainty that the logic of individuals, as well as of social groups, is inconsistent. It has been practically established that any sufficiently large databases are inconsistent too. Contradictory information may be acquired by way of different channels (or even by one and the same channel). (Tabakov, *Logics*)

Paraconsistent logic is goal-oriented differently from classical logic, and this makes it improper to extrapolate the classical aims in the field of paraconsistency. This fact should not detract from the merit of the paraconsistent system; on the contrary it should be considered an asset from a philosophical point of view. They have their place among non-classical logics, which are not closed to intuitionistic logic but rather to modal or many valued logic. Therefore we should regard them in the way we do the above-mentioned logics. There are hardly any efforts to use modal logic as an underlying logic for mathematical theory, but this is no reason to underestimate them, for no one sees their proper use in this. The same holds for paraconsistent logic. The purpose of paraconsistent logic is to account for a kind of epistemological situation, for the logic of a certain social or computer system, perhaps to explicate the logic of a theory, which studies a specific sort of ontological entities. Since paraconsistent logic is not the alternative to classical logic, it is not necessary to construct them as systems, which serve as bases for the mathematical theory framed. Finally, I should like to dwell on the relation between paraconsistent logic and "dialectical logic". Unfortunately, despite frequent references the term "dialectical logic" is not yet clear enough. Neither is the problem of its essence, topics and tasks. In the literature often "dialectical logic" is just a synonym for "dialectics". Dialectical logic must be the logic of dialectics. From this point of view modal logics are closer to dialectical logic. The notions of necessity and possibility are between basic notions of dialectics. Some other kind of logic namely tense, many-valued logic, and those with descriptions are more nearly connected with dialectical logic then paraconsistent logic. As for the so-called "dialectical contradictions", in my opinion, it is not a logical notion. The same holds for the notion of negation used in dialectics. It is close to the notion of opposition and development.

In my view, paraconsistent logic is best understood as a stage of the sequence mentioned above. Otherwise, the dialectical and philosophical importance of paraconsistent logic is overvalued and all other above-mentioned micro-revolutions are understated, especially modal and many-valued logic and their relation to dialectics.

f) Abandonment of monotony - non-monotonic logic (Bell, *Non-Monotonicity*). A logic has been monotonic in the following sense: If we have A|-p and A⊂B, then we also have B |- p. The monotonic character of traditional logic does not depend on the details of the logical system; it is quite fundamental. Usually, much human reasoning is monotonic, but some important human common-sense reasoning is not, especially in legal theory and practice.

I accept the possibility for several more micro-revolutions to take place but my impression is that all basic ideas for taboo breaking have already been tried. I think that there is also a revolutionary aspect in using the apparatus, the methods, approaches and structures of contemporary logic for the analysis of objects of other fields. This is why I would also set this stage aside: studies that put less emphasis on logical inference than on the logical analysis of language.

Usually, problems in knowledge come from borderline cases. The development of science involves specifying, particularizing, and giving shape to and displacement of boundaries. In the development of 20th century logic, such a displacement has usually led to enlargement. Every micro-revolution involves such a displacement and enlargement, and one of the main moments has been the enlargement of the boundaries of the set of logical terms. Syllogistic has been completed by conjunction and disjunction, traditional logic by implication, multiple relations, and quantifiers. Consequently, classical logic has been enlarged by adding modalities or some new kind of negation or implication. For some time, many-valued and intuitionistic logics have been on the borderline, and then modal logic. However, after the emergence of the semantics of possible worlds, the borders have been enlarged and modal logic has steadily come into the interior and even close to the core of contemporary logic. In my view, deontic and especially paraconsistent logics are currently close to the border but on the inside, while the border itself contains non-monotonic logic, erotetic logic and most of the many recently emerging and still not established logical systems.

2. THE LOGIC OF THE 21ST CENTURY – TOWARDS PROLIFERATION OR "NEOFUNDAMENTALISM"

Philosophical challenges of the second revolution.

The non-classical logics have already found their due place, and in the course of time we have been able to overcome the shallow interpretations. A lot of empirical material has been gathered, and the problem is now ripe to be re-interpreted in a new philosophical-methodological light. The methodology of modern logic is falling behind the development of science, and the development of methodology of modern logic is behind that of logic itself. For example, in the correlation ontology - mathematics - logic, the pendulum has shifted from Leibniz's thesis about the ontological universality of science ("logical truths are valid in all possible worlds") towards "logic without ontology". It is important to say that the question about logic is a special one. Logic appears as one entity with two faces: a private and a philosophical science at the same time; a part of mathematics, studying philosophical questions related to truth, thought, and cognition. That is why falling behind is so undesirable. The philosophical-methodological analyses of non-classical logics are considerably behind the specific technical researches. The transition from the classical to non-classical logic raised many problems prior to philosophy of logic. The problems relating to the philosophical interpretation of non-classical logical systems have been assuming an ever-increasing actuality in modern logic. What should be the aim in constructing such systems? What is the position of a given system among the rest of the logical systems? What is their role in the theory of knowledge? In current practice methodological developments lag considerably behind the formal logical investigations. It happens that new systems are being constructed and studied by mathematically well prepared, but methodologically naïve authors. Sometimes the systems are put forward without any philosophical-methodological basis. Often the new systems are accompanied by superficial methodological grounding and “ad hoc” methodology.

An important question from a methodological point of view is how far and to what degree has the appearance of non-classical logic led to the crisis in logic. Does their appearance mark the beginning of a crisis, or are they the symptom of an already existing, latent crisis? In that case, studies of non-classical logic will be rather attempts to overcome this crisis. My stand on this question is that the appearance of non-classical logic is rather a symptom, showing that traditionalistic logic is becoming more and more unable to respond to the requirements conditioned by the development of science and cognition. In this respect the appearance of non-classical logics is one way of trying to overcome the crisis. The way out of the crisis goes through the establishment of a concept of “logicality”, which embraces the non-classical logics (or at least a great deal of them), and especially their confirmation as Logics. Maybe there is not just one crisis, but two. The first one is related to the narrowness and the inadequacy of traditional and classical logic, which led to the appearance of the non-classical logics; the second one is a result of the existence of these logics, which led to proliferation. Common in both cases is a crisis in the foundation of logic, related to such fundamental questions as "the subject and essence of logic and the logical", "the unity of logic, its universality and its boundaries". Logic has been subdivided into a number of logics; so which one now is the “proper Logic”? And can we talk about “proper Logic” at all? A number of alternatives to classical logic have been offered. Questions have been asked such as, "What is logic after all?", "Is logic two-valued or many-valued?". For some non-classical logic’s, their very status as “logic” is reasonably questioned. Is there, after all, a reason why all those disciplines use one and the same word: "logic"? Could we say that some of them are 'more' logical than others, or that they are closer to the 'real' logic? Is every one of those logics a separate, independent discipline on its own, with no connection to the rest, or can they be grouped together in some way? And if yes, then by what criteria? Which ones of them, and to what degree, study the most general laws of thinking, the correlation between thoughts of truth, the laws and the processes of scientific cognition and of cognition in general? Which ones of them and to what degree are related to the question of truth? Which of the non-classical logics are a sort of play of symbols, or a mathematical experiment, which has no relevance (or has only a remote relevance) to Logic; and then, which ones are logics whose "logicality" can be proved by ontological and Epistemological arguments? Even if some of the logics are ruled out as 'illegitimate' logics or 'mathematical exercises with symbols', and still others are ruled out because they employ the word logic only metaphorically, still, for the ones that remain, all those questions are nonetheless valid. All these questions are essential to the philosophical interpretation of modern logic.

The conception of the One Logic used to allow viewing it as a necessary connection of thoughts according to their truth value, valid for all real and possible worlds and theories. However, logic becomes parcelled; many logics have appeared, but which one is the one Logic? Is there anything that allows us to use the same word for all of them? Are any of them more 'logical' than others? Is such logic an independent discipline without a link to the rest, or they can still be grouped together, and if they can, then - how, by what properties? Which of them and to what extent study the most general laws of thought, the connection of thoughts according to their truth, the laws and process of scientific knowledge and knowledge in general? Which of them and how are they related to the question of truth? Which of them are rather games with symbols, mathematical experiments bearing only a distant relationship to Logic, and which are logics whose logicity can be demonstrated by ontological and epistemological arguments? How does the question now stand of the *a priori* nature of logical laws? The philosophy of science calls this subdivision proliferation of the theory, and considers it to be a symptom and a reason for the methodological crisis in the field. Is proliferation in logic the beginning of a crisis or is a sign of a crisis in the development of logic? How is this related to the conceptions of monism and pluralism of logic?

The idea of "logical neofundamentalism"

There are here two positions that have often been confused.

One is 'logical monism', which insists on the uniqueness of logic: 'There is one logic and everything else called "logic" is just an application of this unique logic in one or another field'. The thesis that there is only one Logic often leads to the thesis that this is the traditional (or classical) logic.

The other is 'logical fundamentalism' - the desire to find an undoubtled fundament to logic, to demarcate logic and make it independent, to show what unites the different logics - 'the many logics are types of logic, and, being such, they must necessarily have something in common'. The task to find and demonstrate what is the 'common' factor belongs to philosophy.

A radical possible way out of the methodological situation in logic is **logical antifundamentalism** - 'the many logics are independent disciplines, each with its own object and methodology, therefore there is no proliferation and no reason to talk of a crisis'. Then no logic can justly declare that it is the 'real logic'. However, the proliferation of logic provokes a natural tendency toward fundamentalism, which can be observed in many philosophers of logic. This is why I propose, as an alternative way out, **logical neofundamentalism** - a conception of the character, object and nature of the logical, viewed in both the epistemological and the ontological aspect. The conception of "logical neofundamentalism" must lend support to orientation in the many logics, introduce some kind of structure, order and organization in them. At the same time, it must demonstrate their unity in their diversity, clearly state the common factor that unites them. It must clearly justify whether, in the presence of so many different logics, there can be universality of logic and, if so, in what sense we are to understand it. Such a ("neofundamentalistic") conception must necessarily involve the following aspects:

1. It must provide a clearly argumented conception of logic, which should be a development of the previously existing conceptions of logic; it must include a view of the nature and objects of logic, it must also encompass non-classical logics or at least a considerable part of them, and serve as a criterion for evaluating which logical systems can be called logics and which cannot. It must also allow for the evaluation of systems that have not yet been constructed.

2. It must provide a view of the universality of logic as understood in the sense of item 1.

3. It must provide some kind of structuring of the many logical systems constructed so far; the criteria for such a structuring must be related to the points provided in 1 and 2. This idea of structuring must come from one common position and allow for development.

I will briefly consider these three points.

A new definition for logic. A new conception is needed for the object of logic and the character of the logical. If one analyses the existing definitions of logic, one can see their incompleteness. The science of correct reasoning; science which describes relationships among propositions in terms of implication, contradiction, contrariety, conversion, etc. (*Webster*) Logic is the systematic study of reasoning that provides standards by which valid reasoning can be recognized. It clarifies the reasoning process and provides a means for analyzing the consistency of basic concepts (Popkin); theory of argument and reasoning (Cumbee) and so on. In my opinion a good definition must include the inference and also the link of logic to language and ontology.

There is a concept according to which logic is basically related to language. However, this limits the object of logic too much, and besides, there is the need to specify exactly which areas of language are objects of logic. Ideas form into statements about the names of objects in a given language. Inference is related to the relation between truth of statements, and it is also revealed in a language. The interrelation between ideas according to their truth defines the relations between statements in such a language. The formal structure of thinking can also be seen as a language. Logic studies some general regularities, but they have to be invariable in relation to something. Logic also investigates structures of language, and so can point out those words that are invariable in logical conclusion. The words that actively take part in the structure of logical conclusion and directly affect it - the "logical terms" - also have to be pointed at. Non-logical words (descriptive terms) are usually names of certain objects or properties. A logically true sentence is one whose truth can be proved only on the bases of its logical form and the meaning of logical forms that take part in it. The "formal structure of thinking" can be reduced to investigating language so that logical terms and their properties are pointed out and analyzed. Inference is also reduced to analyses of logical terms. The same applies to the relation between statements according to their truth, as expressed in language. The question of what defines this relation is a philosophical one.

A synthetic definition is needed that would include the linking of logic to thinking and reasoning as well as to truth and logical inference, as well as its link to language and ontology. I propose: 'logic deals with the objectively ontologically predetermined dependence (relation) between thoughts according to their truth, as expressed in language'. On the basis of this definition, a concept about the nature of the logical, regarded in a epistemological as well as in ontological aspect, is formed.

About the universality of logic. As for the universality of logic, I think that the transition from classical to non-classical logic leads to the idea of replacing the concept of a Logic in general; ('logic of the world') by 'logic of a theory, or of a group of theories (about the world)', and thus the universality of Logic is realized through the universality of logics and the universality of "Logicity". The conception of one Logic allows it to be viewed as a necessary correlation between the thoughts of truth and makes it valid for all real and possible worlds. The concept of a multitude of logics renders this impossible. This narrows the universality of each separate logic. It can only be universal as far as it is a logic, that is, as far as its laws are universal for a particular field or theory. It is clear that a logic should be universal with respect to the world it describes, but then a number of different worlds are possible. 'A world' could even mean a certain aspect of another world. The universality of logic will be realized through the universality of the separate logics, inasmuch as they will be complementing each other. The common thing between them will be "Logicality", that is, the very fact that they are Logics and that they express a necessary correlation between the thoughts of truth valid for their corresponding theory. (Tabakov, *Universality*)

This position allows the discussion of a number of controversial issues. Such is Hilary Putnam's thesis that 'some of the necessary truths of logic can sometimes happen to be false by empirical reasons' (Putnam, *Logic*, p.216). Expressed in this form, this statement is unclear to me. It comes out that 'necessary truths' are not necessary. I think that my position allows for an explanation: the truths in question are 'the necessary truths of classical logic' and therefore the formulation becomes 'empirical reasons can replace classical logic by non-classical logic'. Related to 'empirical reasons', ontology defines logic!

Another unclear thesis in Putnam's conception is that 'logic, in a certain sense, is a natural science' - what precisely is that sense? I perceive that as 'logic has its empirics (a somewhat specific one, to be sure) and it is in this sense that "logic is similar to natural sciences"'.

A classification of non-classical logics.

The enormous number of different non-classical logics allows some philosophers to doubt the possibility to classify them. My own attempts and the analysis of many other attempts to classify logical systems (e.g., "map" in Rescher) have convinced me that a successful orthodox classification is only possible if using internal mathematical, syntactical criteria.

However, such a classification would not give much to the philosophical methodological meaning of those systems. Mathematicians classify their structures according to internal criteria but the philosophical methodological look must evaluate them according to whether they are pure models or correspond to 'objective relations of objective things', and to precisely which ones.

Of course, every logical structure constructed as a mathematical system is a potential model and, depending on its interpretation, may serve for modeling in diverse areas. However, our purpose requires a classification corresponding to the philosophical interpretation and importance of the systems rather than to their formal technical elements. This will not be a classification in the precise sense of the word but a "**typologization**", since a system may have different interpretations and the demarcation will not be strictly exclusive. Even in attempts to classify systems in accordance to their main interpretation, there would again be no orthodox classification. (Tabakov, *Idea*)

I propose the following "typologization" of logics, corresponding to their philosophical evaluation (this is a "typologization" and maybe even a classification of the basic, most important and crucial philosophical methodological interpretations of logical systems):

Logical systems serving as bases for a mathematical theory.

In my opinion belonging here are only classical logic and the Intuitionistic Logic. There some efforts to use modal logic or paraconsistent logic or (some other) as an underlying logic for mathematical theory but none of them have been successful. Of course this is no reason to underestimate them, their proper use is in other types as follows.

Logical systems - attempts at an explication of logical inference. These are logical systems related to attempts at a better explication of logical inference. Logical inference is one of the most important problems of logic. It is often interpreted through implication. However, the paradoxes of material implication demonstrated that it does not express well enough logical sequences, does not provide meaningful relation between the two propositions but only their relation according to their truth. The systems constructed in the Frege-Hilbert tradition have set logic apart from the question of logical sequence. The principle of compatibility is a very strong abstraction: it allows the combination of propositions that have no meaningful link between them. The compatibility of arbitrary propositions independently of their content does not raise so many objections in the case of disjunction and conjunction (they are symmetrical and not related to sequencing), but with implication, the situation is very different, especially if the goal is the explication of logical sequences outside mathematical theories. Classical logic stands closest to the logic used in deductive sciences (especially in mathematics, which uses very strong abstractions) and expresses the argumentative thinking in them, and its language is close to the language of exposition of results of mathematics - a result not so much an idealization of natural language but of the language of deductive sciences and of set-theory, extensional thinking. Classical logic is related to the tradition, coming from Aristotle, of subject-predicate structure of the propositions leading to Boolean classes but based to very strong abstractions. Some important questions related to logical sequencing cannot be formulated solely in terms of 'true' and 'false'. Logical systems created with the express purpose to express sequencing are e.g., Zinoviev's complex logic and naturally relevant logic (see e.g., Anderson and Belnap, *Entailment*).

Logical systems expressing epistemological aspects. Modal, many-valued, paraconsistent logic, and to some extent also tense logic and fuzzy logic can be viewed as logics of particular Epistemological situations. Thus the idea of necessary and contingent truths leads to modal logic.

For a certain period in the past the set of logical problems detached themselves from the epistemological ones. After Boole, the problems pertaining to the formal structure of thinking began to be considered independently from the problems of the theory of knowledge. Thus they were presented as a clearer-cut type to be investigated and traditional logic split into two branches "mathematical (formal) logic" and "dialectical logic" The former, which in principle was operating under the influence of the needs of mathematics, can be found in the works of Frege and in subsequent treatments, while the latter is visible in the works of Hegel's followers. This has given some authors the ground to divide the "logical" into formal-logical and "content-logical". A division like this strikes me as a rather artificial one. It is made on methods rather than in accordance with subject matter or problems. What is "logical" is inseparable. The formal system is mathematical models. With the emergence and development of non-classical logic the two branches linked together once again. Classical logic sprang from the problems inherent in the foundation of mathematics. It is an explication of the logic of mathematics involving thinking in set-theoretical terms but itself resulting from rather strong abstractions. The creation of the majority of non-classical logic has been provoked by logico-philosophical considerations; there was the need for weakening the two strong idealizations underlying classical logic on the one hand and, on the other, the transition in logical research from studying logic in general to the investigation of the logic of a definitive type of theory (theories).

Logical systems related to ontological aspects. There is the conception that ontology determines logic and that some entities in themselves impose logic on the theory studying (describing) them. Its argumentation is related to the idea that some properties of the objects of the micro-world require a non-classical logic for the theories studying them. The thesis, that the objects of the micro-world can be successfully studied (solely?) by theories using a logic where the law of distribution is not valid, has separated as an independent field, the logic of quantum mechanics (in particular, Feyerabend; Reichenbach, Van Fraassen,). I would formulate a 'thesis of Reichenbach, Birkhoff, and von Neumann': 'The mathematical apparatus of quantum mechanics requires a logic different from classical logic', which I interpret: 'The ontology of the objects of quantum mechanics imposes a non-classical logic on the theories studying them.' (Or, in weaker versions, 'the ontology of the objects of quantum mechanics imposes (allows for/also) a non-classical logic for (some of the) theories.) The strongest version is that the problems with the quantum mechanics are one of the first symptoms emerging in the initial approach to genuine non-classical science. In the future, there may come much more drastic situations. In the process of cognition we face again and again new entities and theories about them in which the work with classical logic meets considerable difficulties and inconveniences and will lead to essential complications. For theories studying these entities, we will have to rethink many times the universality of some of the laws of classical logic.

The opposite thesis - 'Things can be done well with classical logic alone' - has its supporters too. However, the main question is whether the theories using classical logic do not meet difficulties and whether they guarantee completeness of description.

It is usual for a theory to be initially developed on a classical-logic basis, but the direct application of this logic is not always the most suitable one. Thus, in practice another logic is resorted to, known or assumed to be an hypothesis, a thesis (Hilbert's thesis), that this logic has a translation into classical logic. In such cases the formal systems are mathematical models of this operator logic. In addition to this instance, as has been demonstrated in paraconsistent logic, the use of non-classical logic could be invoked by ontological or epistemological considerations. Certainly, the logic is determined by the epistemological situations, which could well happen to require a many-value or a paraconsistent logic. Logical systems whose semantics is determined by epistemological elements explicate logic as an instrument. However there are cases when certain entities themselves dictate the kind of logic for their investigating theories. For example, certain properties of the object of the micro-world require non-classical logic for their investigating theories. For theories where truths are divided into "ordinary truths" and "necessary truths", a modal logic would be required. Ontology, theory, and the logic of the theory are mutually correlated. One of the tasks of logic is to investigate the necessary interdependencies in terms of the truthfulness between assertions in general, as well as between the assertions of a given theory or discipline. The explication of this logic into a formal logical system enables (by means of the method of mathematical modeling) its deeper-reaching investigation and the elucidation of some of its non-trivial properties. However, if a given discipline is in difficulty or in crisis, it is one of the tasks of the philosopher-logician to point out the necessity of reconsidering its logic as well and, by way of methodological recommendations, to suggest what kind of logic is to be resorted to, and to what extent it is either occasional or imported, and just how adequate it is for the ontology. Indicative in this case would be the idea that the ontology requires another kind of logic when work with the older one not only leads to a gross complication of the exposition but also impairs the completeness and adequacy of knowledge. In this case philosophy should draw attention to the fact and demonstrate that the crisis was not the result of an error or lack of knowledge in the concrete field, but that is rather the result of incorrect application of logic inconsistent with the ontology of the entity investigated. (Tabakov, *Philosophical*).

Logical systems as (only!) mathematical models - and systems created without philosophical justification: they are interesting at least because they allow analogies with other systems interesting to philosophy. Such logical systems can be viewed as "uninterpreted" abstract languages.

Applied logic - logical systems that do not claim to explicate logic in any general philosophical, epistemological or ontological sense, but aim only to express logical relations involved in a particular field of application, such as artificial intelligence, cognitive science, linguistics, computer science etc. The discipline of 'applied modal logic' is very developed, which shows how rich the possibilities for interpretation of modal logic are, as are algorithmic, dynamic and approximation logic. I predict that other similar logics will develop as well - applied many-valued logic, applied paraconsistent logic, applied constructive logic, applied relevant logic, applied non-monotonous logic, applied deontic logic, applied erotetic logic etc.

The boundary. There are also some borderline cases open to discussion - deontic logic, erotetic logic etc. - they are at the borderline currently because, by some of the criteria, there is a discrepancy with my definition of logic. I put them at the borderline because, in the spirit of 'Jorgensson's dilemma', there are doubts as to how far they are related to and deal with logical inference and it is asked if there can be any logical inference in them at all. For me, they are (in the terminology of Lakatos) 'monsters'. I classify them as quasi-logical, 'logicoid' studies of language.[3] However,

I predict that in them too, there will be works related to logical inference, and they will come within the limits of the logical.

Some results of the classification - Non-classical logics "are not what they are".

A comparative analysis of the classification along internal mathematical criteria and a typologization according to philosophico-methodological criteria must be offered. The question must be developed synthetically - the systems must be classified according to syntactical criteria, they are also typologized according to their philosophical interpretation, and the two types of analysis must be compared. This comparative analysis contributes to understanding the place and the importance of each separate logical system. It shows which philosophical interpretation would correspond to certain syntactic structures, as well as which syntactic structures would best fit the attempts to build up a logical system with a previously set philosophico-methodological aim. Logics in which the two-value principle does not hold are closest to logical systems typologized as logics related to Epistemological situations, and also to the controversial but very interesting type of logics related to non-classical ontologies. For systems typologized as related to logical sequencing, a non-observance of the principle of combinativity and of extensionality is convenient. This approach allows the rethinking of some existing systems. For instance, the historically established unification of the question of the logic of necessity with the question of sequencing (related to the paradoxes of material (implication) is not necessary. The two questions could also be considered separately, with different methods and instruments. Many modal logics do not observe the principle of functionality. However, if we consider that modal logics are logical systems related to Epistemological situations, them modal operators should transform in a certain manner the truth value of propositions into another truth value. Then, it is reasonable to also construct modal logics in which to retain in some form the principle of functionality, probably some kind of many-valued functionality. The systems of Lewis are built upon classical logic, but would not the idea of a logic with a different implication be realized successfully in systems constructed differently? I think that the logics of possible worlds are philosophically important enough that they can be separated as an independent type. Unlike extensional interpretations, here the truth of a proposition depends not solely on the truth of A in the particular world but also on its truth in other possible worlds. Thus possible worlds are viewed in themselves as two-valued but with a notion of truth relative to the particular possible world. This is an elegant and beautiful generalization of the classical two-valued logic: it is the logic of a universe with only one possible world. The necessity to take into account the character and degree of truth of the separate propositions of one area leads to multiple possible worlds in it, which leads to a non-classical logic for their universe! After the emergence of the semantics of possible worlds, modal logic can be defined, rationalized and interpreted as a logic of possible worlds.

The idea to construct a relevant logic as a subsystem of classical logic is similar to the idea of Intuitionistic logic (another subsystem). However, the question stays open whether the philosophical ideas on which they are based would not be better realized in logical systems with a qualitatively different type. Classical logic is so constructed as to be a convenient basic logic for mathematical theory. How far can methods suitable for one kind of questions be convenient in resolving quite different types of questions? The typologization helps to distinguish between logics explicating logical inference, logics of possible worlds and logics of necessity and contingency. The modal systems of Lewis are closest to the second kind. Fuzzy logics are not so closely applied to questions related to epistemology and dialectics where often 'fuzzy' and 'rough' terms are used. For me, applied logics are those in which the specific nature of the particular field imposes certain requirements on the language, methods of study and the operations considered. The best example is the operation of iteration in dynamic logic.

Figuratively, some of the systems of modern logic justify the expression: "**Non-classical logics are not what they are.**"

The place where a system is located in the course of a typologization essentially determines the criteria and methods to evaluate this situation. For systems which, in typologization, are evaluated as pure mathematical models, the basic feature is the mathematical perfection of the system: precision, correctness and interesting theorems. When the system is typologized as a logical system serving as the basis for a mathematical theory, the above requirements are also completed by the requirement for it to be appropriate for such a basis. In applied logical systems, it is important that the system is a successful model related to the particular questions of the specific field to which the system is directed. For all of these types, the priority is to one or another extent the mathematical perfection of the system. When, however, the system is typologized from philosophical positions, e.g., in logical systems aiming at a better explication of logical sequencing, or in logics related to certain Epistemological situations, or logics related to new non-standard ontologies, the situation is radically different. Such a typologication considerably increases the philosophical methodological importance of the system, but, at the same time, it increases the philosophical methodological requirements for it. When the basic value of the system is mathematical, the criteria are mathematical and the judges are mathematicians. When the system is closely related to philosophical methodological questions and conceptions, when it explicates notions and categories of philosophical nature, the criteria for it should be philosophical and its judges should be philosophers! Then, arguments of the kind of 'the system is not interesting because it does not contain any interesting (mathematical!) results (theorems)' are no longer valid – it is not interesting from the mathematical perspective but it may have a great philosophical importance.

The transition from the classical to non-classical logic and the transition to non-classical science

The transition from classical to non-classical logic is related to the transition from classical to non-classical (and post-neoclassical) science. Science as a type of rationality and the rationality itself of scientific knowledge have been, until the early 20th century, related to classical logic. To a great extent, however, classical logic has exhausted its heuristic and methodological potential. There is already talk of non-classical science, non-classical ontology. The empirics of classical science is the middle structural level. The theories of the classical science are the empirics that have led to the creation of classical logic. They do indeed also provide a model for studying borderline cases, the mega-world and the micro-world, but it is controversial how far such an extrapolation can be relevant. Classical science involves notions like locality, causality, observability, stability and stationary objects, while the non-classical involves illocality, indeterminism, unobservability, changeability, fractality and non-stationary objects. Elements of non-classical science are observed in quantum mechanics, synergetics, cosmogony, genetics, cosmology, and some cognitive sciences. Classical logic is related to statements of classical science and it is not *a priori* evident that it can be extrapolated to contemporary non-classical and post-non-classical science. With such fields, it may be more appropriate to use, e.g., modal logic, many-valued logic, non-monotonous logic, quantum logic, maybe paraconsistent logic, and why not also intuitionistic logic or some logic with a strong negation. The objects of non-classical science sometimes provoke a crisis in the corresponding field, leading to new theories. The logic of these new theories is precisely the new object for logical study.

There is often talk of non-classical rationality; for me non-classical "logicity" is an extension of rationality. The limitation of classical rationality in science is related to the limitation of classical logic, the essence of human rationality cannot be reduced to or based solely on the utterly rigorous laws of classical logic.

Indeed, the theory of sets works with qualitatively new, considerably more complex terms and objects: is it not one of the first manifestations of non-classical science? This thesis may be surprising, the set theory has established itself very well into the foundations of mathematics and is thus perceived as something 'classical'. However, if we remember the definition of non-classical science – 'qualitatively new, considerably more complex notions', is this enormous hierarchy of infinities then a sign that something is involved here that could well satisfy a definition of non-classical science? Non-classical science is related to the objects of the macro- and micro-world, but are the objects and constructions in contemporary mathematics only the less complex ? Are the monstrous sets not comparable to the objects of the macro-world, and the infinitely small ones to the objects of the micro-world?

The language of the philosophy is closer to the language of non-classical logic than to that of the classical. Non-classical logics involve philosophical methodological considerations; this is why they can be justly deemed philosophical logics. Often the working mathematician identifies the concepts of logic and the logic of mathematics, for him, this is *the* Logic. However, non-classical logics are not related to the logic of mathematics, they do not serve as the basis for a mathematical theory (with the exception of intuitionistic logic), this is why for a very long time non-classical logics were not interesting to mathematicians. The big boom in the establishment of non-classical logics came when it was demonstrated that they also have an essentially applied character, e.g., in fields like the computer science. This is an important but not fundamental argument, since my basic thesis is the philosophical methodological importance of non-classical logic related to issues of epistemology, dialectics and ontology.

NOTES

[1] The transition to non-classical logic has often been compared to the transition from Euclidean to non-Euclidean geometry but this is superficial and imprecise: the analogy is done by external signs while the essential question is left in the background, namely, what sphere precisely is concerned, and that is Logic! The science studying the most general laws of thinking; the science studying the correlation of thoughts by their truth value; a sphere related to the laws and processes of cognition and knowledge, and to the question of truth. However important geometry may be, it is only one of the disciplines of mathematics. In its relation to the universal philosophical concepts, to the world, to being, to the 'logos' and 'ontos', it cannot be compared to logic.

[2] I do not have in mind the many non-classical logics that are not related to mathematics: if I had dealt with them, I would not have written so 'easily': many-valued and generally non-classical logic is almost never used in mathematics. I speak here about some extensions of the first-order classical predicate logic, about higher-order logics, as well as about some versions of constructive logics, which only have an episodic use in contemporary mathematics.

[3] In the spirit of the terminology of Lakatos, I would call this method 'the method of isolation and fencing in monsters'. A new cell is constructed – a borderline area where the monster is closed in. Thus it 'stays' but at the same time it is 'closed in' and harmless. How far this is an *ad hoc* solution, is another question.

[4] And in some sense classical logic is also "not what it is"!

REFERENCES

Anderson, A. and N. Belnap. *Entailment: The Logic of relevance and necessity*. Princeton: Princeton Univ. Press,1975.

Asenjo, F. A "Calculus of Antinomies." *NDJ of Formal Logic*, V.7 1966: 38-49.

Asenjo, F., J.Tamburino. "Logic of Antinomies." *NDJ of Formal Logic*, V16, 1975: 56-71..

Bell, J. "The logic of Non-Monotonicity." *Artificial Intelligence*, V.41(1990): 206-215.

Birkhoff G. and von Neumann, J. 'The Logic of Quantum Mechanics". *Annals of Math.*, 37(1936): 111-134..

Cohen, I. *Revolution in Science*. Cambridge: Cambridge Univ. Press, 1985.

Da Costa, N. "On the Theory of Inconsistent Formal Systems." *NDJ of Formal Logic*, V.15 1974: 225-239.

Feyerabend, P. "Reichenbach's Interpretation of Quantum Mechanics." *The Logico-algebraic Approach to Quantum Mechanics*. Dordrecht, 1975.

Hintikka, J. and G. Sandu. "A Revolution in Logic?" *Nordic Journal of Philosophical Logic*, 1996, Vol.1, No.2: 169-183.

Jaskowsky, S. "Propositional Calculus for Contradictory Deductive Systems". *Sudia Logica* ,. XXIV(1969): 83-97.

Petrov, S. "Hegel's Thesis of Contradictory Truths." *Intern. Logic review*, 17-18 (1978): 91-104.

Polikarov, A. *Revolutions in Physics*. (in Bulgarian) Sofia: Academic Publishers, 1996.

Putnam, H. "Is Logic Empirical?" in R.Cohen and M.Wartofsky (eds.), *Boston Studies in the Philosophy of Science* (Proc. of the Boston Coll. 1966-1968), vol. V, Dordrecht: Reidel, 1969: 216-242.

Tabakov, M. "Goedel's theorem in retrospect." *Bull. of the Section of Logic*, PAS, V(1984): 66-72.

Tabakov, M. "Logics tolerant to contradictions." *Ruch filozoficny*, 13(1989): 13-28.

Tabakov, M. "Philosophical aspects of non-classical logical systems." *Logic and Applications*, Vol.III. (1992): 131-135.

Tabakov, M. "Philosophical importance of paraconsistent logics" *Abstracts of the Int. Congress of Logic, Methodology and Philosophy of Science*. Moscow, 1987, V.5.

Tabakov, M. "Universality of Logic or Uniqueness of the Separate Logic" *Universal and Particular* Sofia,1999:125-128.

Tabakov, M. "The Constructive Logic or Logic of Intuitionistic Mathematics." (in Bulgarian) *Filosofski Forum* 1 (1999): 122-139.

Tabakov, M. "An Idea About a Classification of the Logical Systems." (in Bulgarian) *Filosofski Alternativi*. Sofia 1 (1999): 46-62.

Vakarelov, D. "Notes of N-lattices and Constructive Logic with Strong Negation." *Studia Logica*, 1-2(1977), 36-42.

Van Fraassen. "The Labyrinth of Quantum Logics." In R.S.Cohen and M. Wartofsky (eds.), *Boston Studies in Philosophy of Science*, Vol. 13, Dordrecht: Reidel, 1974: 224-254.

Part IV

PHILOSOPHY OF SCIENCE AND COGNITIVE SCIENCE

DAFINA GENOVA

IDEALIZED COGNITIVE MODELS AND OTHER MENTAL REPRESENTATIONS

This paper provides arguments for and against cognitive linguistics as an emergent paradigm in the theory of language. Idealized cognitive models, postulated by some proponents of cognitive linguistics, are compared to other mental representations such as prototypes, archetypes and stereotypes. The old controversy for realism or cognitivism in science, respectively in linguistics, is again brought to the fore . The author is in favor of a dualistic approach to the study of language, since reference and conceptualization presuppose each other – in order to refer, we have to conceptualize, and we cannot conceptualize, unless we refer.

The two fundamental questions that any theory of language has to explain are: "How does language relate to the world?" and "How does language relate to the mind?" The answers to these questions are not easy because they are dependent on how philosophers and psychologists answer another fundamental question: "How does the mind relate to the world?" According to the analytical school in philosophy, words relate to objects in the world; we can have beliefs and desires about things in the world and the propositions that sentences express can be true or false. The cognitive approach to language focuses on how the mind conceptualizes our experiences of the world as a result of its cognitive abilities, making use of psychological notions such as gestalt, figure and ground, frame, attention, salience, etc.

The adherents of analytical philosophy seem to assign a greater role to language and more specifically to a theory of meaning or semantics than the proponents of cognitive linguistics. Analytical philosophers believe that it is only by the analysis of language that we can analyze thought:

> Thoughts differ from all else that is said to be among the contents of the mind in being wholly communicable: it is the essence of thought that I can convey to you, the very thought I have, as opposed to being able to tell you merely something about what my thought is like. It is of the essence of thought, not merely to be communicable, but to be communicable, without residue, by means of language. In order to understand thought, it is necessary therefore, to comprehend the means by which thought is expressed (Dummett , *Analytical*, 195).

Studying the semantics of natural language will help, analytical philosophers believe, to answer philosophical questions about the reality represented in thought, philosophical questions about reference and truth, for it is through language that we express the general and the particular, what is true or false and what is an assertion,

Dimitri Ginev (ed.), Bulgarian Studies in the Philosophy of Science, 129-140.

a question or a command. Language serves many purposes. It is used to express a logical truth, a deduction as in (1):

(1) All men are mortal.
Socrates is a man.
Therefore, Socrates is mortal.

an analytical truth, as in (2):

(2) Bachelors are unmarried.

and a synthetic or a contingent truth, as in (3):

(3) The car is red.

We use language, among other things, to describe a state of affairs in the world, to express a mental state, to perform a linguistic act or to speak about its own aspects when we use it as metalanguage.

Semantics as a component of natural language is important to philosophers because it is meaning or intension that picks out a referent in the world (excluding indexical expressions). The philosophical approach is focused on thought; the cognitive one on the way concepts form hierarchies, clusters and configurations and the impact of the latter on the structure and organization of language. According to the philosophical approach, language is regarded not only as a means of expressing thought, but also as a topic that deserves attention in its own right, which is not the case with the cognitive approach. Sentences of natural language are seen as a function of the meaning of the words used and the rules of syntax. They are the result of the principle of compositionality, bigger units consist of smaller units. In more abstract terms, following Frege, a sentence is the result of the completion of a predicate expression, e.g., ( ) *is writing*. When we complete it, we get the sentence *Peter is writing*. Reference and predication (Searle, *Speech*) illustrate, in yet another way, the same principle. We use linguistic expressions to refer and to predicate. We use words (words themselves do not refer) and other linguistic expressions to refer to an object in the world, e.g., *car,* and we use words and other linguistic expressions to assign a property to an object, e.g., *red*, and the result of reference and predication is an assertion, *The car is red.*

The most serious argument that proponents of cognitive linguistics provide in its favor is figurative language and the fact that the "logical" approach to language does not give it due attention:

> Figurative language is generally ignored in current theories; at best it is handled by special, ad hoc descriptive devices. Yet it would be hard to find anything more pervasive and fundamental in language, even (I maintain) in the domain of grammatical structure (Langacker, *Foundations, I,* 1).

> human thought processes are largely metaphorical. This is what we mean when we say that the human conceptual system is metaphorically structured and defined. Metaphors

> as linguistic expressions are possible precisely because there are metaphors in a person's conceptual system (Lakoff and Johnson, *Metaphors*, 6).

After claims about the "pervasive" nature of metaphor and after detailed comments on various types of conceptual metaphors like "structural", "orientational", "ontological" and "conventional metaphors" and after explaining metaphor as a "mapping from a source domain to a target domain" (Lakoff, *Women*), all that we learn about metaphor (a term burdened with many senses even before the cognitive paradigm in linguistics came into fashion) is that it is partial understanding of concepts in terms of other concepts:

> We claim that most of our normal conceptual system is metaphorically structured; most concepts are partially understood in terms of other concepts (Lakoff and Johnson, *Metaphors*, 56).

It also becomes clear that we can "understand a sentence such as "'Inflation has gone up' as true"(Lakoff and Johnson, *Metaphors*, 170) and that "an understanding of truth in terms of metaphorical projection is not essentially different from an understanding of truth in terms of non-metaphorical projection" (ibid., 197). In other words, Lakoff and Johnson acknowledge the truth of *"Snow is white is true if and only if p*", postulated in truth-conditional semantics, on statements, interpreting p as "a statement in some universally applicable logical language" (ibid., 183). In the standard literature, p is equivalent to "truth-conditions" and "statement" is equivalent to "the use of a given sentence on different occasions", and uses of sentences usually imply context, and context is usually not a part of a logical language.

Understandably, Lakoff and Johnson, being ardent proponents of cognitive linguistics, avoid using the term "truth-conditions" because by it is meant "the circumstances in the world that have to hold for a sentence to be true", i.e., sentences through truth-conditions are related to the world, and as is well-known, cognitive linguists ignore not only the relation between language and the world, but the world as objective reality itself (Lakoff and Johnson, *Metaphors*, 184; Jackendoff, *Languages*, ch. 8). For Lakoff and Johnson, truth is based on understanding one concept in terms of another, i.e., truth is "relative to a conceptual system" (Lakoff and Johnson, *Metaphors*, 180).

It has to be mentioned that understanding is not a notion that only proponents of cognitive linguistics believe. For example, another current version of "Snow is white is true if and only if p" in truth-conditional semantics is: "To know the meaning of a sentence is to know its truth-conditions." And does not knowing presuppose understanding at least in some cases? I think we can all agree that it does. Understanding in analytical philosophy has a different connotation (in the philosophical sense of the term). It means, e.g., to

> have an implicit understanding of the working of language, that is, of the principles governing the use of language; it is these principles which relate to what is open to view in the employment of language, unaided by any supposed contact between mind and mind other than via the medium of language that endows our sentences with the senses that they carry (Dummett, *Analytical*, 195).

Similarly, Lakoff and Johnson do not use the term "proposition" (or thought, at least in one of the senses of the term) because propositions or thoughts for Frege and some other philosophers are not psychological but abstract entities:

> By a thought I understand not the subjective performer of thinking, but its objective content, which is capable of being the common property of several thinkers (Frege, *Sense*, 52).

Arguments for the "legitimacy" of intentional notions, e.g., propositions, is given in Moravcsik (*Thought*, ch. 1), one of them being that different sentences express the same proposition ("being the common property of several thinkers"), for example, *I love you*, *Ich liebe dich* and *Обичам те*.

By laying a stress primarily on the ability to understand, cognitivism in linguistics seems to ignore other abilities of our cognitive system that are based on understanding: the ability to reason; practical reasoning of the form in (4):

(4) X wants to acquire Y
In order to acquire Y, X must do A
Therefore, X does A and acquires Y

and theoretical (and logical) reasoning, where one comes to a conclusion on the basis of certain premises, abilities that are also represented in language.

Some proponents of cognitive linguistics postulate that concepts form hierarchies, clusters and configurations; for analytical philosophers it is beliefs that are interrelated, a fact acknowledged by psychologists, too. Here are some examples:

> The formation of belief involves a preference or ranking among alternative projections of what will be, must be, or ought to be. In coming to belief p we normally come to favour p over alternatives q, r, etc. Furthermore, since there is no guarantee that the favoured belief is true, all beliefs should be construed as projections of reality, not copies of it. Beliefs can be understood prior to their assessment as true or false (p, q and r are propositions, the comment is mine) (Moravcsik, *Thought*, 98).

> Conscious experience in particular are those which are *avalable to acts of thinking which are reflexively available to further thinkings*. Conscious occurrent thinkings – conscious acts of wondering-whether, judging-that, supposing-that and the like – are those which are *made available for further, indefinitely reflexive thinking* (Carruthers, *Language*, 194).

> one belief demands many beliefs, and beliefs demand other basic attitudes such as intentions, desires and... the gift of tongues (Davidson, *Rational*, 473).

In order to master the cognitive paradigm in linguistics, we have, among other things, to learn that a **predicate** is the "semantic pole" of a morpheme (Langacker, *Foundations, I*, 97); that a **predication** is the "semantic pole" of any linguistic expression (ibid., 97); that **meaning** is "cognitive processing" (ibid., 5); that a **domain** is "contexts for the characterization of a semantic unit" (ibid., 147); that **imagery** is "our ability to construe a conceived situation in many different ways (seeing it from different perspectives, emphasizing certain facets over others, approaching it at different levels of abstraction, and so on)" (ibid., 47) (the ordinary sense of **images** being "representations of appearances" (Carruthers, *Language*, 32)); that **context** is only "the cognitive representation of the interaction between cognitive categories" (Ungerer and Schmid, *Introduction*, 54); that a **cognitive model** is "the sum of the experienced and stored contexts for a certain field by an individual" (ibid., 55); that an **event** is "a cognitive occurrence of any degree of complexity, be it the firing of a single neuron or a massive happening of intricate structure and large-scale architecture" (Langacker, *Foundations, I*, 100), not to mention again what is understood by **metaphor** (Lakoff and Johnson, *Metaphors*, 56). And since cognitive grammar "claims that lexicon, morphology and syntax form a continuum of symbolic units serving to structure conceptual context" (Langacker, *Foundations, I*, 35), only few of the traditional linguistic terms (units) are mentioned, e.g., **morpheme** as in *stapler* (verb + *er*) and **linguistic expression**, e.g., *responsible for* (ibid., 35) and *Your football is under the table* (ibid., 279). Mentioning **word** and **sentence** is carefully avoided. New terms replace familiar ones, e.g., **profile** and **base** replace **figure** and **ground** (ibid., 183).

It is not only the case that a different sense is introduced for familiar terms, but it is also the case that many adherents of cognitive linguistics are not consistent with the terminology they have introduced. For example, linguists using **semantic marker** as a theoretical construct in word semantics use it also for the analysis of sentence meaning (not very common, but possible, e.g., Ravin (*Lexical*)) and one would expect (at least I would), proponents of cognitive linguistics to do the same. Instead, when it comes to analyzing sentences, they do not use the notion **cognitive model**, but a notion already in use, e.g., **semantic roles** as in Langacker (*Foundations, I*, 190) or the notions **noun phrase** and **verb phrase** as in Deane (*Grammar*). Also, it is not always clear what the relation of cognitive models and more specifically of **idealized cognitive models** (ICM in Lakoff's sense (*Women*)) and the components of language is. Vervaeke and Green are very explicit in this respect:

> Vandeloise (1990, p. 412) notes that ICMs are linked by Lakoff to such diverse entities as particular words, modifying phrases (such as "strictly speaking"), mental processes such as vision, scenarios, live individuals, and proportions. Thus, we find Lakoff offering an ICM for the Pope on one page (Lakoff, 1987, p. 71) and another for the proposition expressed by the sentence, "please, sit in the apple juice", on another (Lakoff, 1987, p. 148)". (Vervaeke and Green 1997, http://www.yorku.ca/faculty/academic/christo/papers, p. 11).

It might be a brilliant idea to characterize "semantic structure" "relative to knowledge structures" obliterating the distinctions between metaphorical and non-metaphorical "projection" of concepts; linguistic expressions, after all, are symbols of something, but that something is not only concepts. Cognitive processing does not necessarily result in endless chains of concepts in which each concept is understood in terms of another. There are smaller and bigger chunks of configurations of concepts, e.g., *to be responsible for* and *Your football is under the table* (Langacker's examples quoted earlier). If there are sentences on the linguistic level, there must be some entity that corresponds to sentences on the mental level.

The mental entity that *Your football is under the table* stands for is a proposition, and there are psychologists who are not ashamed of using the term (Steinberg, *Psycholinguistics*; Anderson, *Architecture*; Johnson-Laird, *Mental*; Stevenson, *Language* and others) having in mind by it, of course, its most general sense – a thought. One of the arguments in favor of the propositional representation of sentences is the fact that not only phrases, e.g., *bread and butter*, but also simple sentences, e.g., *Mary had a lamb* seem to be represented in memory "in their entirety" (Steinberg, *Psycholinguistics*, 122) and not as separate constituents, e.g., concepts. Another argument in favor of the same type of representation is that thoughts can occur spontaneously in the form of "inner verbalization", again as whole units.

By denying that propositions are mental entities, some proponents of cognitive linguistics also seem to be denying that human reasoning is based on inference rules, mental models and heuristic schemas and that they operate on propositions and not on concepts. It is obvious that we cannot separate categorization (and conceptualization in the sense explicated here) from reasoning. Recent experimental findings in psychology show that inductive probabilistic reasoning, among other things, involves "projection" of properties of a category to another similar term, e.g., *Dogs can get disease* is used as a premise for inferring *Cats can get disease* (Heit, *Knowledge*). In other words, the cognitive approach in linguistics seems not to be explicit on the difference between **understanding** on the one hand, and **propositional thinking** and **reasoning with propositions**, on the other, and how this is formulated in language.

Language surely seems to be more than mere symbols for the understanding of one concept in terms of another. Language is "involved" in conscious propositional thinking (Carruthers, *Language*, 134). Even Steinberg, who is one of the proponents of the view that thought is independent of language, acknowledges that:

> language could be said to affect the content and direction of particular thoughts. These particular important instances are: (1) language may be used to provide new ideas; (2) language may be used to bring about a change in beliefs and values (Steinberg, *Psycholinguistics*, 116).

Our competence of "the working of language" can persuade us, for instance, after reading a book on Buddhism, to change our religious beliefs. Also, prior linguistic competence is required for the acquisition of many concepts and beliefs (Carruthers, *Language*, 85). For example, one cannot imagine a human being not speaking any language to entertain thoughts about, say, the structure of a tumor cell. As far as images are concerned, they are not excluded from the process of thinking (ibid., 36). We can entertain a "mixed thought" (mixed with images) when reasoning about some everyday practical problem, e.g., hanging a picture on the wall. We can think *I can put it...,* the inner verbalization being followed by an image of the wall.

Some proponents of cognitive linguistics have a neurophysiological understanding of the notions **thought** and **concept**, e.g., Langacker:

> what I call a thought is the occurrence of a complex neurological, ultimately electrochemical event, and to say that I have formed a concept is merely to note that a particular pattern of neurological activity has become established, so that functionally

equivalent events can be evoked and repeated with relative ease (Langacker, *Foundations, I*, 100).

According to the philosophical position, thoughts occur (and we are aware of their occurrence as opposed to animals, which are capable of primitive unconscious thinking only); they cause the occurrence of further thoughts; they have content and are formulated in language (Carruthers, *Language*, 229).

By denying any dualism about the structure and organization of the human mind and by placing our experiences mainly in perception (What about acquiring knowledge indirectly, through inferencing?) human understanding can be seen to a large extent as an automatic process, and because of this the "formation, testing and changing of beliefs" are not seen as "key mental processes" (Moravcsik, *Thought*, 48) that involve volition. A good counter-argument to the view that "a thought is [not only] the occurrence of an electrochemical event" is our conscious awareness of various cognitive abilities like **knowing**, **thinking**, **believing**, **understanding**, **planning** and **explaining** that are expressed in sentences of natural language: *I don't know it, I think you are right, I can't believe you did it, I think you understand*, etc. Proponents of cognitive linguistics are also expected to answer another question: "If understanding is taken as an unanalyzable primitive, what is its relation to **interpreting**, **integrating**, **assessing** and **presuming** – cognitive practices that we employ in everyday conversation"?

Some proponents of cognitive linguistics claim, as we have already mentioned, that conceptualization is "grounded" in experience (no-one would deny that) and their account of spacial prepositions is convincing from that point of view (Jackendoff and Landau, *Spacial*, 99-124; Lakoff, *Women*), but I am just curious how they would explain conjunctions standing for logical concepts like *and*, *or*, *if...then* and *not* in terms of experience (no such account has yet been offered as far as I am concerned and for obvious reasons). I am also most curious how *The question that arises then is to what extent one is justified in attributing to the composite structure any of the properties that distinguish these various possibilities from one another* (Langacker, *Foundations, I*, 281); a sentence from theoretical discourse is constructed from concepts understood in terms of other concepts and, more specifically, how any of the postulated cognitive schemas THE CONTAINER, THE PART – WHOLE, THE CENTER – PERIPHERY or THE SOURCE – PATH – GOAL (Lakoff 1987) is exemplified in it. Or, how a sentence from a metalanguage, e.g., *Nouns have gender, case and number* can be analyzed in similar terms. In other words, if abstract human thought is "grounded" in concrete experience and if metaphor is a basic mode of human cognition (Lakoff and Johnson, *Metaphors*), that is, if metaphor relates abstract experiences to salient aspects of our concrete experience of the world, how are abstract structures like rules, principles and laws explained in terms of conceptual metaphor? Another question that arises is: "If "nominal expressions" like *mind*, *thought*, *concept*, *perception*, etc. are understood only as "convenient reifications" (Langacker, *Foundations, I*, 100), what is the nature of primitive cognitive representations like **space**, **time**, **vision** and **color** that some proponents of cognitive linguistics postulate as "basic domains" in their cognitive linguistics"? Are they innate concepts (to use the more accepted term) as some psychologists and philosophers claim? Are they acquired or are they only inherited predispositions? If our understanding is grounded in perception and if cognitive schemas are pervasive, why is it that very often we realize that we do not fully understand what other people say? I agree that the reason for misunderstanding is internal but the reason seems to be what some scholars call intentionality as a general property of the mind.

One can also question the "irreducibility of basic domains". For Keller (*Review*) "mappings" from a "source domain" onto a "target domain" are simply analogies that

> are only possible as a consequence of inherent similarities in a given source and a given target domain. The paradox lies in the fact that an analogy is not irreducible if it relies on a priori structural similarities. What might inherent similarities be, if not yet prefigurative (dare I say literal?) structural representations? (Keller, *Review*, 777).

In other words, the similarity is preexistent and not established after "mappings" between domains.

For many proponents of cognitive linguistics, understanding is based on structural configurations of concepts; philosophers speak of "conceptual dependency", the only difference being in the point of view of the two paradigms. The starting point in cognitive linguistics seems to be mind-to-language while the starting point in analytical philosophy seems to be language-to-mind-to-reality.

For some cognitive linguists it is not only the case that new senses of familiar terms are introduced; it is also the case that terms overlap or coincide with other well-established terms. For instance, it is not clear what the difference is between **conceptualization** and a **concept** (for arguments that Lakoff (*Women*) does not make such a distinction, see Vervaeke and Green (http://www.yorku.ca/faculty/ academic/ christo/papers)). Also, there seems to be some similarity between an ICM and an **archetype**, a similarity acknowledged by Langacker himself:

> Numerous constructs are characterized with reference to certain "idealized cognitive models" (Lakoff 1987) that approach the status of archetypes, being grounded in everyday experience and fundamental to our conception of the world (Langacker, *Foundations, II*, 13).

Langacker is probably right to claim that archetypes and cognitive models have something in common. They both are **idealized representations of experience**, but this is where the similarity ends. In the case of archetypes, typical human experience is attributed to represent "superhuman, or even cosmic significance", while no such claims are made about cognitive models. Archetypes are "unconscious psychic dispositions" for creating myths and for reacting emotionally; they also link the conscious and the unconscious. Archetypes are at the basis of self-awareness and self-knowledge, and in this sense, they seem to have a more restricted significance compared to claims made about cognitive models. The images on the conscious level that point back to a very few basic archetypes are of a limitless variety and are most often not under conscious control. For that reason, we may fear and deny them and try to repress them. Obviously, cognitive models do not have such a fateful significance.

Stereotypes are also idealized mental representations (they can be inaccurate) of, at least, natural kinds of objects, and they have to be compared rather to **prototypes** than to ICMs because stereotypical representations are representations of real world knowledge and not of the imposition of subjective features of an object in itself, as it is the case with ICMs. When commenting on stereotypes, we shall have in mind Putnam's understanding of the term that he uses in the explication of word meaning on the epistemological level. For Putnam

> In ordinary parlance a 'stereotype' is a conventional (frequently malicious) idea (which may be wildly inaccurate) of what X looks like or acts like or is (Putnam, *Meaning*, 249).

Stereotypes are conventional beliefs (like *Tigers are striped*) that can be also "based on untypical or unrepresentative members of a natural kind" (ibid., 250) ("three-legged tigers and albino tigers" are still tigers) as distinguished from prototypes, which are the best example of a category (e.g., a sparrow is a prototypical bird while penguins are not). Most often though, stereotypes "capture features possessed by paradigmatic members" of a class (ibid., 250) and in this respect they are not much different from prototypes. The difference between the two is that with stereotypes as beliefs they are socially determined in the same way as extension. For example, in American culture "speakers are required to know what tigers look like", while in American Indian culture speakers are required to be able to tell an elm from a beech (ibid., 249) and in this sense stereotypes are similar to ICMs, which are also culture specific.

One can look for and find other similarities between the mental representations and other mental representations that many cognitive linguists are guided by. Mental representations, no matter how different they are in other respects, have one aspect in common – they are idealizations (or generalizations) of the entity they represent (a real world, mental or a linguistic one), e.g., no-one has seen a prototypical bird or the archetype of a woman, the anima. We are not going to look for all the possible similarities and differences among mental representations. What is of great importance is the stance cognitive linguists take to the representation of linguistic rules – an issue that we shall come back to later and comment on in greater detail.

To sum up for now, many cognitive linguists reject the classical model of categories (the necessary and sufficient features) and replace it with the prototype theory; by claiming that language is highly metaphorical, they also reject truth-conditional semantics, which relates language to the world (in model-theoretic terms, of course) and by positing the view that grammar is only symbolic of the "construal of conceptual content" they reject the autonomy of syntax (rejected, as is well known, by other linguists, philosophers and psychologists for various reasons). Generative grammar is also rejected altogether since it is viewed as "a full and precise mathematical characterization" of the well-formed sentences in a language.

As has been mentioned, many cognitive linguists are against formalization (for them Chomsky's approach is also formal, not to mention Tarski's or Montague's); they are in favor of postulating the least possible number of principles in a theory – a doctrine to which they are not the only adherents. For instance, the Minimalist Program is based on "the sparest set of assumptions"; relevance theory, a cognitively grounded theory of linguistic communication (Sperber and Wilson, *Relevance*) is based on a single principle – the principle of relevance – although, one has to admit that if sparsity of principles is always theoretically justifiable, it might not always capture the essence of the entity/ies studied. What is more important is that "syntactocentric" theories are not psychologically plausible – when we produce utterances we do not start from deep structure (syntax) and get to the surface structure (semantics). Semantically based grammars, as is well known, are more suitable for the explication of the two basic language abilities – the production and understanding of utterances and from that point of view cognitive linguistics is more acceptable, since it does not attribute secondary significance to meaning. The point is that meaning is equated with conceptualization and cognitive processing, and as a result, language is seen not "as a direct object of inquiry", but only as a means of explicating how concepts are understood. Linguistic rules and structures are not granted an independent status; they are grounded in general cognitive schemas and cannot be separated from their instantiation (Langacker, *Foundations, II*, 535) and yet it is not made obvious how such claims are substantiated. At least to some, rules and principles are metarepresentations of preexisting observable structures.

A major objection to the cognitive paradigm in linguistics is the fact that cognitive schemas are only hypothesized without being grounded in empirical results the way that prototypes and prototype effects are from experiments in cognitive psychology. Or, to quote Ackerman:

> It is one thing to provide evidence for how inferring, metaphorical mapping, etc. are accomplished by means of language use, but quite another to demonstrate that constraints on grammatical structures, or on linguistic systems quite generally derive from conceptual structures (Ackerman, *Observations*, 377).

He is also right to claim that the term "conceptual" is more appropriate than the term "cognitive" in this context since the latter is the more general of the two and would thus subsume generative grammar and other cognitively grounded linguistic theories. In the sense used, "cognitive" rather means "conceptualization" as a crucial explanatory component of grammar" (ibid., 366).

Linguistic expressions are, to remind the reader again, only symbols of conceptualization. As a result the traditional distinction between the lexicon, morphology and syntax is obliterated and linguistic rules are rejected. The closest one can get to linguistic rules are cognitive schemas and in this respect cognitive linguistics is very close to connectionist modeling of cognitive processing (Rumelhart and McClelland, *Parallel*; Rumelhart et al., *Schemata*) in cognitive science where no distinction is made between general and specific knowledge and where no rules are postulated. The model is parallel and distributed and is based on excitatory links between elements within a language constituent and inhibitory links between elements in separate constituents. In connectionist simulation of cognitive processing the mind/brain does not contain a central processing unit as is the case with the modular account of cognitive processing. The brain system is locally managed in terms of trial and error, and it has the ability to modify and shape itself employing prototype knowledge representations. Paradoxically, one of the arguments of cognitive linguists for the rejection of generative grammar is its "formalism", and yet at the same time cognitive linguists embrace connectionist modeling, which is worse in this respect – it is more abstract and it abounds in mathematical formulas.

The attempts of cognitive linguists to place linguistic theory in a wider context are not isolated. Chomsky himself holds the view that the study of language has to be a branch of cognitive psychology and that it is the task of psycholinguists and psychologists to specify the details of a performance model. Similarly, Sperber and Wilson's theory of linguistic communication is also grounded in cognitive psychology. Undoubtedly, further empirical research not only in cognitive psychology, but also in cognitive science in general, will help us learn more about the cognitive basis of language. The point is that a linguistic theory has to combine a cognitive conception of language with a communicative one, since very often we utter a sentence expressing a certain thought in order to communicate something different (excluding metaphor, metonymy and irony). In other words, cognitive linguistics explicates private but not public language. Even Chomsky has admitted (Chomsky, *New*) that language is not only "individual" and "internal" but "external" as well. It is another matter that he holds the view that the latter cannot be defined. For example, when we try to be polite, we are not guided by psychological considerations alone, we are guided by social ones as well, that is, in linguistic analysis different factors have to be taken into consideration, e.g., the social aspects of verbal interaction. In this respect cognitive linguistics can only complement but not replace theories that regard grammar as the product of more than one factor.

I have provided arguments for and against cognitive linguistics. I personally am not in favor of the methodological determinism that is typical of some theories that maximize a doctrine or a point of view while minimizing other possibilities, and many presentations of cognitive linguistics are such examples. After all, one has always to bear in mind that if syntax is not autonomous, language is. It is its own means and devices that we put to use when we communicate.

REFERENCES

Ackerman, Farrell. "Some observations on the typology of linguistic theories." *Journal of Linguistics* 36.2 (July, 2000): 365-381.

Anderson, John R. *The Architecture of Cognition*. Cambridge, Massachusetts: Harvard University Press, 1983.

Carruthers, Peter. *Language, Thought and Consciousness*. Cambridge: Cambridge University Press, 1996.

Chomsky, Noam. *New Horizons in the Study of Language and Mind*. Cambridge: Cambridge University Press, 2000.

Davidson, Donald. "Rational Animals." Ernest Lepore and Brian McLaughlin (eds.) *Actions and Events. Perspectives on the Philosophy of Donald Davidson*. Oxford: Oxford University Press, 1982.

Deane, Paul D. *Grammar in Mind and Brain: Explorations in Cognitive Syntax*. Mouton, de Gruyter, 1992.

Dummett, Michael. "Can Analytical Philosophy Be Systematic and Ought It to Be?" Kenneth Baynes, James Bohman and Thomas. McCarthy (eds.) *After Philosophy: End or Translation*? Massachusetts: Massachusetts Institute of Technology Press, 1987, 185-215.

Frege, Gottlob. "Sense and Reference." *Philosophical Writings*. Paul Geach, and Max Black (ed. and trans.) New York: Philosophical Library, 1892/1956, 56-78.

Heit, Evan. "Knowledge and concept learning." Koen Lambert and David Shanks (eds.) *Knowledge, Concepts and Categories*. London: Psychology Press, 1997, 7-41.

Jackendoff, Ray. *Languages of the Mind.* A Bradford Book. Cambridge: Massachusetts, London: England, The Massachusetts Institute of Technology Press, 1992.

Jackendoff, Ray, and Barbara Landau. "Spacial language and spacial cognition." Jackendoff, Ray. *Languages of the Mind.* A Bradford Book. Cambridge: Massachusetts, London: England, The Massachusetts Institute of Technology Press, 1992, 99-124.

Johnson-Laird, Philip N. *Mental Models*. Cambridge: Massachusetts, Harvard University Press, 1983.

Keller, Janet. "Review of Johnson 1987." *Language* 64.4 (1998): 775-777.

Lakoff, George. *Women, Fire and Dangerous Things*. Chicago: University of Chicago Press, 1987.

Lakoff, George, and Mark Johnson. *Metaphors We Live By*. Chicago and London: University of Chicago Press, 1987.

Langacker, Ronald W. *Foundations of Cognitive Grammar*. Vol. I. Stanford: Stanford University Press, 1987/1999.

Langacker, Ronald W. *Foundations of Cognitive Grammar*. Vol. II. Stanford: Stanford University Press, 1991/1999.

Moravcsik, Julius M. *Thought and Language*. London: Routledge, 1990.

Putnam, Hilary. "The meaning of meaning." *Mind, Language and Reality*. Cambridge: Cambridge University Press, 1975/1989, 215-271.

Ravin, Yael. *Lexical Semantics without Thematic Roles*. Oxford: Oxford University Press, 1990.

Rumelhart, David E., and James L. McClelland (eds.) *Parallel Distributed Processing. Vol. I: Foundations*. Cambridge, Massachusetts: The Massachusetts Institute of Technology Press, 1986.

Rumelhart, David E., Paul Smolensky, James L. McClelland, and Geoffrey E. Hinton *Schemata and Sequential Thought Processes in Parallel Distributed Processing: Explorations in the Microstructure of Cognition.* Vol. II, Cambridge, Massachusetts: The Massachusetts Institute of Technology Press, 1986, 7-58.

Searle, John R. *Speech Acts. An Essay in the Philosophy of Language*. Cambridge: Cambridge University Press, 1969/1987.

Sperber, Deirdre, and Dan Wilson. *Relevance: Communication and Cognition*. Oxford: Oxford University Press, 1986.

Steinberg, Danny D. *Psycholinguistics. Language, Mind and World*. London and New York: Longman, 1982.

Stevenson, Rosemary J. *Language, Thought and Representation*. Chichester: John Wiley and Sons, 1993.

Ungerer, Friedrich, and Hans-Jorg Schmid. *An Introduction to Cognitive Linguistics*. London: Longman, 1996.

Vandeloise, Claude. "Representations, prototypes and centrality." Savas L. Tsohatzidis (ed.) *Meaning and Prototypes*. London and New York: Routledge, 1990, 403- 437.

Vervaeke, John, and Christopher D. Green. "Women, fire and dangerous theories." *Metaphor and Symbol* 12 (1997): 59-80. Quoted in: http:// www. ca/faculty/academic/christo/papers/.

LILIA GUROVA

PHILOSOPHY OF SCIENCE MEETS COGNITIVE SCIENCE: THE CATEGORIZATION DEBATE

A standard story about what happened with our views on concepts and categorization is broadly accepted in cognitive science circles. The crucial role in this story is played by the notion of the so-called *classical view* as far as all other views have been conceived as either alternatives, or improvements of that view. The notion of the classical view that supports the standard story is based on three main assumptions: (1) The classical view is a descriptive theory: it is about what concepts are, not about what concepts should be. (2) The classical view presupposes that concepts are mental representations and its main claims are about the structure of the conceptual mental representations. (3) The basic statement of the classical view is that concepts are represented by means of a set of individually necessary and jointly sufficient defining features. This paper will show that the three assumptions listed above are not true for most of the theories to which the notion of the classical view is intended to refer. As a result, the standard story is a source of unsound criticism having far-reaching negative effects: underestimation of the status of well-defined concepts in science, misunderstanding of the relations between the classical theories of concepts and categorization and their alleged alternatives.

1.INTRODUCTION

A standard story about what happened to our views on concepts and categorization is broadly accepted in cognitive science circles. According to this story, since the times of Aristotle until recently, i.e., until the second half of 20th century, the Western mind has been occupied by the view – most often called the *classical* view – that concepts and categories are represented by means of defining features, which may be regarded as individually necessary and jointly sufficient conditions for something to be considered an instance of a particular concept or a member of a particular category. The standard story holds that although the classical view has always had its critics, it was only seriously challenged for the first time in the 1950s by Ludwig Wittgenstein in his *Philosophical Investigations*. Wittgenstein demonstrated that the instances of most, if not all concepts, in fact do not possess features that are shared by all of them[1]. In empirical science, the challenge to the classical view appeared some twenty years later (in the 1970s) in the form of the experimental findings of Eleanor Rosch and her followers. As a result of these challenges, another view on concepts and categorization has been established that

Dimitri Ginev (ed.), Bulgarian Studies in the Philosophy of Science, 141-162.

today is broadly known as the *prototype* view[2]. Very soon, however, a lot of people working in psychology, linguistics and philosophy realized that the prototype view is not without its own serious flaws. This realization has inspired part of them to look for alternatives both to the classical and to the prototype view. Their efforts ended in theories known as the *exemplar* view (Smith and Medin) and the *theory-based* view[3] (Murphy and Medin). Another part of the research has been directed to the improvement of the so-called classical view or to the search for a synthesis of the best sides of the existing rival theories of concepts and categorization. This research produced a series of *neo-classical* theories of concepts. A recent collection of papers edited by Lawrence and Margolis (1999) is a good illustration of the latest version of the standard story.

What this paper would like to draw attention to, is that the standard story, as sketched above, about the change in views on concepts and categorization, is not simply a neutral chronological summary of past ideas, study results and their interpretations. Being itself a kind of interpretation, the standard story imposes serious constraints on the vision of what actually has already been done and what could be, as well as what should be, done further in order to obtain to a better understanding of concepts and categorization. There have been some critics of this story, questioning, for example, the claim that the transition from the classical to the prototype view should be considered as a progressive shift (Margolis). The aim of this paper is to look at the standard story from a different perspective. There is a notion that plays a crucial role in that story – the notion named *the classical view* on concepts and categorization. Its role is crucial, since in the framework of the standard story the classical view is like a frame of reference: all other views have been regarded as either alternatives or improvements on that view. This paper will show that the way the standard story deals with the classical view is based on three erroneous assumptions. The first assumption is that *all* theories to which the classical view is intended to refer (i.e., *all* philosophical, logical, psychological and linguistic views on concepts and categorization that have ever been in circulation since the times of Plato and Aristotle until Wittgenstein's *Philosophical Investigations)* are descriptive ones. The whole criticism against the classical view, which the standard story keeps tracks of, does not take into account the normative character of some of the theories to which the classical view is intended to refer[4]. The great difference between descriptive theories and normative theories is that the former address questions like "What *are* concepts?" or "What is *to have* a concept?" while the latter are interested in the question "What *should be* a concept?". But a criticism that is valid for a theory whose main claim is that "All concepts *do have* clear-cut boundaries" is not necessarily valid for a theory that claims that "All concepts *should have* clear-cut boundaries". In this way the neglect of the normative character of certain views might become a source of unsound criticism. In what follows it will be shown that this kind of unsound criticism not only takes place in discussions dominated by the standard story, but it also produces far-reaching negative effects, including a criticism of the status of well-defined concepts in science (or at least in some sciences such as biology).

The second assumption that the standard story associates with the classical view is that concepts are mental representations and that the main thesis of the classical view is about the structure of the conceptual representation. This assumption also leads to misinterpretations of certain experimental findings, as will be shown in part four of this paper.

The third controversial assumption that the standard story associates with the classical view is that all statements constituting "the hard core" of the classical view are implications of one central statement that concepts have individually necessary and jointly sufficient defining features. Most critics of the classical view, for example, tend to regard the statement

(1) Concepts have clear-cut boundaries.

as a simple implication of the alleged central statement

(2) Concepts have individually necessary and jointly sufficient defining features.

That (2) implies (1) is without any doubt true. No-one, however, asks the questions whether (1) could be true without having (2) and whether (1) does not play a more important role than (2) in those theories categorized as classical. A careful analysis (but the standard story lacks it) should convince us that we could have clear-cut boundaries without having a list of defining features. In other words, we could have a well-defined concept (a concept with clear-cut boundaries) by having some defining *rule,* which should not necessarily be a conjunction of defining features. The role of a defining rule could be played by a theory, for example. In this way the theory-based view on concepts does not seem so much an alternative to the classical view as the standard story claims. The defining rule view on concepts could be regarded as a generalization both to the classical and to the theory-based view. Respectively, the classical view and the theory-based view are to be conceived as particular instances of the defining rule view. This conclusion, however, could not possibly be drawn by one who accepts the assumption made by the standard story that the central statement of the so-called classical view is the claim that concepts have individually necessary and jointly sufficient defining features.

It would be naïve to think that the standard story that I discuss in this paper is nothing but an occasionally accepted wrong reconstruction. Ironically enough, this story is a result of the operation of the same principles that every theory of categorization would like to discover, i.e., the principles that lead a particular category to be viewed as distinct as possible from all other categories. These principles urge everyone who proposes a new theory of concepts and categorization to represent their views as contrarily as possible to all previous views concerning the same matter. In this way, their views become more salient on the basis of all other views. At the same time, however, they lose the perspective on how a more powerful theory combining the strong points of all previous views could be created.

In these introductory notes, some terminological conventions will also be discussed. Up to this moment the terms "concept" and "category" have been used as if there has been a well-established consensus about their meanings. Unfortunately, this is not the case. I do not have in mind only the well-known difference between the common philosophical and the psychological use of the term "category"[5]. The situation is much more complicated because even in psychological and cognitive science circles there is no agreement on what should be considered a concept and what a category. This lack of terminological consensus is, on the one hand, part of the problem that has initiated the debate about concepts and categorization and, therefore, one should not expect it to be solved before coming to a theory of concepts and categorization conceived as satisfactory enough by the whole scientific community. On the other hand, any study in this direction must start with some preliminary assumptions concerning the basic terminology. Represented in this way, the situation looks a bit paradoxical, but we could break the paradox accepting that any preliminary terminological notions could be changed at any stage of the study if this turns out to be necessary. Nevertheless, one should be careful enough with one's choice of preliminary definitions, for despite the stipulations about their provisional character, they could still prejudice the whole of the work that follows.

In brief, it will be accepted in this paper that a category is a group of things (never mind what kind of things – real or imagined, physical or mental) that are considered to be equivalent (Rosch, *Principles,* 191), and that a concept is something which refers to, or represents, a category (this definition however does not imply that concepts are necessarily *mental* representations).

In what follows, first, the birth and the growth of what is called here the standard story will be traced briefly. The aim of this historical reconstruction is to show how the notion of the classical view appears as a theoretical construct, the meaning of which is a function of a particular theoretical framework, and how this notion has shaped the discussions about the nature of concepts and categorization. The third part of the paper provides some evidence against the three main assumptions of the standard story. In the fourth part the main arguments against the classical view are discussed. The aim here is first, to show how the erroneous assumptions of the standard story lead to unsound criticism of the classical view, and second, to reveal the negative consequences of this criticism. Finally, in the concluding remarks, a prospect for a possible reconciliation of the views represented by the standard story is outlined.

2. A BRIEF SKETCH OF THE STANDARD STORY: THE INVENTION OF "THE CLASSICAL THEORY OF CONCEPTS" AND THE BEGINNING OF THE CATEGORIZATION DEBATE

The standard story claims that the categorization debate in cognitive science has been inspired by the work of Eleanor Rosch. The interesting point here is that the way the standard story represents Rosch's findings is surprisingly different from her own interpretations. As this discrepancy seems to be important for the discussion undertaken here, let us regard it in more detail.

2.1. Rosch's principles of categorization

Rosch's views on categorization are summarized in her 1978 paper *Principles of categorization*. For Rosch, categories are groups of objects which, for certain reasons, are considered equivalent (Rosch, *Principles*, 191). At the very beginning of the paper she points out that her theory is only about *the formation of categories in culture* (i.e., about how different cultures decide on considering some objects to be equivalent in some respect); it is not about *the development of categories in children* (about how children learn to treat some objects as equivalent), nor is it about *how categories are processed in the minds of adult speakers of a language* (about how an adult decides whether a particular object belongs to a particular category) (Rosch, *Principles*, 189-190). As we shall see, this warning has often been neglected by most of her contemporaries, as well as by almost all of her later readers. Briefly stated, the answer that Rosch gives to the question of how categories are formed in cultures is the following: human categorization does not seem to be

> an arbitrary product of historical accident or of whimsy but rather the result of psychological principles of categorization, which are subject to investigation. (Rosch, *Principles*, 189)

According to Rosch, there are two basic principles of categorization. The first principle, Rosch calls it "the principle of cognitive economy", holds that the main function of category systems is

> to provide maximum information with the least cognitive effort. (Rosch, *Principles*, 190).

The second principle asserts that

> the perceived world comes as structured information rather than as arbitrary or unpredictable attributes. (Rosch, *Principles*, 190)

In order to understand how Rosch came to the formulation of these principles, we have to go back and to look into the way she has been involved in the study of categorization. One of Rosch's professors at Harvard was Roger Brown. In the 1950s, Brown, working together with his graduate student Eric Lenneberg, carried out a series of experiments intended to test the Sapir-Whorf hypothesis claiming, that the particular language one has learned as a mother tongue produces significant effects on cognitive processes including perception, memory and thinking. What Brown and Lenneberg obtained (see their later review) was weak support for the Sapir-Whorf hypothesis, or they found out that most people tend to recognize more easily those color hues that are "codable", i.e., for which they can easily find names for. For various reasons, Brown and Lenneberg made their experiments with English native speakers only. In the 1970s E. Rosch had the chance to visit the Dani tribe in New Guinea. What makes the Dani interesting is the fact that their color coding system is completely different from that accepted in the English speaking world. Dani have only two names for colors – *mola* for bright, warm hues, and *mili* for dark, cold ones. Rosch replicated (with some modifications) the experiments of Brown and Lenneberg and found that despite the fact that Dani's codable hues are quite different from the codable hues of English speaking people, the results they produced in the recognition test were very similar to the results of the English native speakers. She concludes from here that it is the non-verbal visual color categorization that precedes, and partially determines, the conceptual categorization, not the opposite, as the Sapir-Whorf hypothesis predicts. It is not difficult to see in this conclusion the predecessor of Rosch's second principle of categorization which holds that the world appears to us already structured and that this perceptual categorization precedes and in some way constrains our conceptual and linguistic categorization.

What about the first principle claiming that the main function of category systems is to "provide maximum information with the least cognitive effort"? Here it is also not difficult to recognize the influence of Rosch's earlier experience with the categorization of colors. In her 1978 paper, Rosch takes for granted that

> most, if not all, categories do not have clear-cut boundaries. (Rosch, *Principles*, 196).

(Remember that, for Rosch, categories are groups of objects that are, for certain reasons, considered equivalent.) In other words, the world appears to be structured to us but there are no clear-cut boundaries given. This claim might be questionable if generalized for the whole of human experience, but it is completely true for the way the world of colors is given to human beings with normal color vision. It is a well-established fact that the spectrum of colors is continuous and that the boundaries between the colors in this spectrum are, from an objective point of view, arbitrary. The function of Rosch's first principle of categorization is to explain exactly how and why we cut up the continuous world in more or less definite categories. In order to think about that world, and to be able to communicate our thoughts to other people, we need a reasonable number of more or less stable categories mapping the real world structure, which are at the same time as informative as possible. That is why the principle of cognitive economy

> dictates that categories tend to be viewed as being as separate from each other and as clear-cut as possible. (Rosch, *Principles*, 196)

According to Rosch, until recently, only one way of achieving this is known - "by means of formal, necessary and sufficient criteria for category membership". She stresses that

> the attempt to impose such criteria on categories marks virtually all definitions in the tradition of Western reason. (Rosch, *Principles*, 196)

Her main claim, however, is that there is another way to make the continuous categories distinguishable: it is by conceiving them in terms of their clear cases rather than in terms of their boundaries (Rosch, *Principles*, 196). Rosch recognizes Wittgenstein as the first to grasp this possibility.

It is important to point out when discussing Rosch's views on categorization that Rosch does not regard the two ways of describing categories (by definitions and by their clear examples) *as incompatible, or rival models.* She does not suggest that the defining features view of categorization is a wrong view, which should be replaced by "the best examples view". The controversy between both views, so typical for the standard story, was created later. Another very important point, which deserves attention, is that Rosch speaks about different *ways* one could *view* the fuzzy categories. She does not speak about different *mental representations.* I think, it is this slightly behaviorist stance that has kept Rosch away from the wrong descriptivist assumptions and generalizations of the standard story.

In short, despite the great attention and acknowledgment that Rosch's findings have received, these findings have been significantly reinterpreted in at least three very important points for the theory of categorization.

First, as Rosch herself points out, her view concerns only the formation of categories in cultures, it should not be regarded as a theory of category learning, nor as a theory of how adults categorize things using already established categories. Many of Rosch's followers, however, have been tempted to extend the scope of validity of her main claims far beyond the area of cultural categorization. Second, for Rosch the well-known defining features view (which will later be named "the classical view") and the view she introduces as "the best examples view" are nothing but descriptions of two different ways people use to conceive categories "as separate from each other and as clear-cut as possible". Therefore, both views are equally legitimate. According to Rosch, the great flaw of all previous thinkers is not that they have supported a wrong view but rather that they have been convinced that this view refers to the only way people categorize the world. The followers of Rosch, however, transformed this into the claim that in fact we have two different theories or views of categories and categorization and that the old one (the defining features view) is wrong, and the new one (the prototype view proposed by Rosch) is the better theory of categorization. Third, Rosch does not regard the opposition between the defining features view and the best examples view in terms of mental representations. She has never addressed the questions of how categories are *represented* in human minds. Later, as we shall see, the whole discussion has been shifted inside the mental representation paradigm and, those who did that forgot completely that the defining features view did not necessarily imply a theory of mental representation.

The standard story, however, is not critical, or even sensitive, to the reformulations listed above. On the contrary, it contributes to their construction and propagation inside the scientific community. Surprisingly enough, the standard story in fact hides the original theory of categorization proposed by Rosch and puts in its place a rational reconstruction, which in fact constitutes a quite different theory. The latter is regarded as an alternative to another very controversial theoretical construct – the so-called "classical theory of categorization". Let us see how and for what reasons this has happened.

2.2 The invention of "the classical theory of concepts"

One has to be cautious when claiming that certain persons have been the first to have introduced something important, but in the case of the standard story of the categorization debate we could hold with a high degree of certainty that it is the 1981 book of Smith and Medin that lays the foundation of the story.

In 1981 Smith and Medin declare as their primary task to systematize the main empirical findings which have given support to the belief that

> the view of concepts inherited from Aristotle was severely lacking. (Smith and Medin, vii)

and that this view should be replaced by a new one. They also declare that the main opposition between the "old" and the "new" views on categorization is already outlined due to the efforts of Eleanor Rosch and her collaborators, but that there are still some problems, which need further clarification. One of these problems is, according to Smith and Medin, the lack of agreement on

> exactly which findings impugned Aristotle's notion of concepts. (Smith and Medin, vii)

It is natural to assume that if one would like to reassess the main arguments against a certain view, one has first to clarify of what this view exactly consists. Smith and Medin do follow this way of reasoning. They start with an attempt to give an account of the old view of concepts, inherited from Aristotle, which they call "the classical view". According to their preliminary definition, this is the view holding that all instances of a concept share common properties and that these common properties are individually necessary and jointly sufficient to define the concept (Smith and Medin, 1)[6].

At first glance, there is no crucial difference between Smith and Medin's definition of the classical view and what Rosch considers as the defining features view. The only difference seems to be that Smith and Medin are more radical in their assessment of that view: while Rosch claims that the defining features view is *not the only way* to view categories as separately as possible but it is still *a legitimate way* to do it, Smith and Medin suggest that *the legitimacy of the view is under question.*

In fact, however, the difference is rather more serious. Rosch claims that the defining features view is nothing but an instrument that people use to make categories as separate as possible, Smith and Medin insist that "the classical view" is a theory of concepts, i.e., theory giving answers to the questions "what concepts are" and "how are concepts represented", they say directly that "the classical view is a theory about representations". According to Smith and Medin, all views on concepts and categorization could be understood in the framework of two basic questions: (1) Is there any single or unitary description (representation) for all members of a given category? and (2) Are the properties specified in the unitary description (representation) true of all members of the category? (Smith and Medin, 3). Smith and Medin argue that the classical view says "yes" to both questions, what they call "the probabilistic view", i.e., the view that, according to them, corresponds to

Rosch's best examples view and which claims that "instances of a concept vary in the degree to which they represent the concept, and consequently vary in the degree to which they represent the concept" (Smith and Medin, 2), says "yes" to the first but "no" to the second question, and their own view, which they call "the exemplar view", i.e., the view that

> there is no single representation of an entire class or concept, but only specific representations of the class's exemplars. (Smith and Medin, 2)

and which they regard as even more radical in its criticism of the classical view than the probabilistic view, says "no" to the first question ignoring in this way the second one.

Later on we shall see how the proposed formal account of the "classical view" has influenced the argumentation against that view, but before going to this most important point, let us look at what happened to the standard story after Smith and Medin set it in motion.

Maybe here is the place to make an important note. One should not be left with the impression that Smith and Medin created the standard story from nothing. This is not the idea that I would like to suggest. Closer to the truth is the claim that they just expressed the psychological Zeitgeist of the 1970s. Smith and Medin were not even the most radical critics of the "old view". They, for example, make the remark that one should be very careful when interpreting experimental findings as direct arguments against the classical view. All experimental findings, say Smith and Medin, are connected with the process of categorization (the process of classifying things by adults using already known categories and concepts). As a theory of representation, however, the classical view does not imply any statements concerning processing that could be evaluated against the experimental findings. Therefore, one should add some processing assumptions to the classical theory. Since different kinds of processing assumptions are in fact compatible with the classical view, one could produce different classical *models,* which could be tested by experiments. Therefore, when some experiments are regarded as arguments against the classical view, one should take into account that in fact the arguments have to be conceived as directed against the particular classical model, not against the classical theory as a whole. This comment, however, does not significantly change the spirit of Smith and Medin's book which, as I already mentioned, should be conceived as a manifestation of the *Zeitgeist* of the 1970s. The essence of this spirit could be briefly expressed in the following way: until recently a certain, let's call it the classical view, has dominated philosophical and psychological thought, it has been shown however that this view has serious defects and it has to be improved or replaced by new, better views. Smith and Medin hold that in order to be successful in doing that, one should first identify exactly of what the main defects of the classical view consist. The hypothesis they propose is that the most serious flaw of the classical view is its claim that every concept has an unitary description, which is true for all instances of the concept and which is not true for all non-instances of that concept. Later on, however, other diagnoses of the main defects of the classical view appeared. What it is important to stress here is that neither the definition of the classical view nor the diagnosis of its main flaw are based on analysis of any real instances of that view, on the contrary, both the definition and the diagnosis were drawn from a set of theoretical assumptions (every theory of concepts is a theory of representation), in other words, the notion of the classical view was introduced as a theoretical construct.

2.3. The development of the standard story

Four years later, Gregory Murphy and Douglas Medin, one of the authors of Smith and Medin's 1981 book, come to another diagnosis of the main defect of the classical view. It is, according to them, that this view assumes that category formation is basically similarity-based. Medin and Murphy claim that this assumption does not help us to understand why some categories are more stable than others. In fact, we could imagine a lot of groups of objects that share some properties: the set of black ink pencils used by eight-year-old girls, the set of all hard cover books stored in my bookcase, the set of red roofed houses built after 1989, and so on. No one of these groups, however, forms a stable category or a concept while at the same time some other groups, whose members do not look similar, have been viewed as stable categories. Medin and Murphy's hypothesis is that the defining-features view could not explain this fact because it does not take into account both the internal structure of the representation of a given category and its external theoretical and explanatory relations. According to Medin and Murphy, not only the classical view but also its alleged rival – the prototype view – suffers from the same disability. That's why both views should be replaced by a new, theory-based, or explanation-based view of concepts and categorization. So, the most important question that separates the rival theories of concepts and categorization is not whether there is any unitary description that is valid for all instances of a given category, rather it is the question about *the structure* of the conceptual representation. According to the classical view, concepts are represented by means of lists of independent defining features, the prototype view claims that concepts possess prototypical structure, while in the end the explanation-based view holds that concepts are organized around a particular theory.

It is important to stress here that although Murphy and Medin seem to reformulate the standard story proposed four years earlier by Smith and Medin, they do not make any significant change to its basic assumptions that all theories of concepts are (and should be) about how concepts are represented and that the central statement of the so-called classical view is that concepts have defining features.

Another interesting point is that Medin and Murphy do not pay any significant attention to the fact that the theory-based view implies clear-cut boundaries between members and nonmembers of a given category, just as the classical view does. The original example that Murphy and Medin discuss is the biblical division of animals into two categories: clean and unclean animals. The group of clean animals includes most fish, grasshoppers, and some locusts, whereas the unclean animals include camels, ostriches, crocodiles, mice, sharks, and eels. What is it that makes the difference between these groups? Murphy and Medin point out that it is not the similarity among the members of any of both categories that determines the conceptual distinction but "some theory or explanatory framework":

> the concepts of clean and unclean animals rests on a theory of how the features of habitat, biological structure, and form of locomotion should be correlated in various animals... Roughly speaking, creatures of the water should have fins and scales, and swim, and creatures of the land should have four legs. If a creature conforms with this theory, then it is considered clean. But any creature that is not equipped for the right kind of locomotion is considered unclean (e.g. ostriches). (Eysenck and Keane, 270)

It is perfectly clear that the "explanatory framework" described above, implies a strong definition of what should be considered a clean animal, and what unclean, although not in the form of a list of independent defining features. And it is also perfectly clear that this definition does not leave room for "unclear cases". At the same time the claim for existence of clear-cut boundaries of concepts and categories has always been regarded as one of the core features of the classical view although it has usually been conceived as an implication of its central statement that concepts possess defining features. It comes out, therefore, that at least from this perspective, the classical view and the explanation-based view do not look crucially different. In other words, if we accept Medin's later version of the standard story (1989) that there have been two important shifts in our views concerning concepts and categorization – first, from the classical to the prototype (or probability) view and second, from the prototype to the theory-based view – the second shift should not be considered as a further move away from the classical view, for it looks like a partial return to the values of clear-cut boundaries. However, as I have already mentioned, this fact has not yet received much attention. We will return to it later. Now let us say what happened with the standard story after the invention of the theory-based view.

A lot of people involved in concepts and categorization studies have taken the question about conceptual structure seriously. Only the proponents of the so-called conceptual atomism (Fodor) claim it to be an inappropriate question because, according to them, concepts, like atoms, have no structure at all. Even conceptual atomists, however, do not escape from the basic assumptions of the standard story, that is why, despite their criticism, they also contribute to its consolidation.

Let us now take a closer look at the evidence against the standard story's basic assumptions.

3. EVIDENCE AGAINST THE THREE MAIN ASSUMPTIONS OF THE STANDARD STORY

In this part of the paper some evidence will be given to support the claim that the three main assumptions of the standard story are not true, at least for some of the theories included under the umbrella of "the classical view". The standard story's main assumptions are: (1) The classical view is a descriptive theory: it is about what concepts are, not about what concepts should be; (2) The classical view presupposes that concepts are mental representations and its main claims are about the structure of the conceptual mental representations; (3) The basic statement of the classical view is that concepts are represented by means of a set of individually necessary and jointly sufficient defining features.

Let us see now why these three assumptions should not be considered true.

3.1. Most classical theories are normative theories: they are not about what concepts are, but rather about what concepts must be

One could easily find a lot of examples supporting this claim, especially examples concerning logical and scientific concepts. Frege, for instance, writes the following

> In all directions these same ideals can be seen at work – rigour of proof, precise delimination of extent of validity, and as a means to this, sharp definition of concepts. (Frege, 1)

Christoph Sigwart, another famous 19th century logician, a representative of the so-called psychologistic trend in logic who has been attacked by Frege and the other proponents of mathematical logic, shared very similar views on this point. He explicitly says that "the concept may be viewed as an ideal" and that it is "the mark at which we aim in our endeavor to attain knowledge" (Sigwart, 246). Sigwart pays attention to the fact that we have different kinds of concepts and that we consider different features to be important for these different kinds. What one should seek in a metaphysical concept, for example, is that the concept must be "an adequate copy of the essence of things" (Sigwart, 246). For logical concepts one only requires them to be "absolutely fixed and determined", there is no need for a logical concept to represent any essence, or even any real thing. Scientific concepts, however, need both to copy the essences of real objects and to be at the same time logically perfect.

Why science needs metaphysically adequate concepts seems more or less clear but why does it need logically perfect concepts? The most obvious answer is: "for the purpose of judgment" (Sigwart, 247). The interesting point here is that later on in the philosophy of science, the claim that *all* scientific concepts should be *individually* metaphysically adequate was subject to debate. The view was proposed that it is enough for the theory as a whole to be proved metaphysically adequate but the concepts that are part of the theory should not necessary correspond to any real thing. At the same time, no serious attack has ever been launched against *the need* for logically perfect, i.e., well-defined concepts. It is true, however, that there have often been discussions about *the possibility* of defining a particular term. Starting from Leibniz's dream of a perfect scientific language, going through the unsuccessful program of three generations of positivists and ending with some recent attempts to re-assess the practice of scientific inference, one could easily grasp that the criteria of metaphysical adequacy have changed several times during the last three centuries but that the norms concerning the determinacy of scientific concepts have not.

3.2. Not all classical theories accept that concepts are mental representations

There are some clear cases of classical theories that definitely do not regard concepts as mental entities and some of them are broadly known. Frege's theory is maybe the best example. In his *Foundations of Arithmetics,* Frege proclaims as one of his main principles

> always to separate sharply the psychological from the logical, the subjective from objective (Frege, X).

Strictly following this principle, he makes a clear distinction between *ideas,* which possess psychological (mental) reality and which are, therefore, something subjective, and *concepts,* which are something completely different – non-mental and, therefore, objective.

The paradox is that at least some of those who support the criticism of classical theory know Frege's view of concepts as abstacta, but the only conclusion they draw from the realization of this fact is that Frege, as well as all other philosophers sharing the same view, are simply wrong when refuting the mental reality of concepts. The critics of the classical view are not aware of what will be shown in the next part of the paper, that most of their criticism would be true only if the classical view presupposes that concepts are mental representations.

Frege is not the only philosopher who insists that concepts should be distinguished from ideas, which are mental entities. In a similar way, Moritz Schlick, for example, holds that

> a concept is to be distinguished from an intuitive image above all by the fact that it is completely determined and has nothing uncertain about it... Thus a concept is not an image. It is not a real mental structure of any sort. It is something that we assume in place of images with strictly determined content. (Schlick, 20)

The most important point here is not whether concepts have or do not have mental reality (whether they are or are not mental representations). The point is whether being a *representation,* either internal (i.e., mental) or external, is the most important, the defining feature of a concept. Not only philosophers such as Frege think about concepts as abstracta, but also some psychologists and logicians of the 19th and the early 20th century have recognized the importance of distinguishing between what they usually call *ideas* (but also images, mental representations, of even psychological concepts), which are "natural psychological productions" (Sigwart, 245) and which are normally more or less vague and ill-determined, and *concepts,* which are products of a rational, goal-oriented human activity, directed toward the satisfaction of certain norms about what the perfect concept should be. The so-called classical view is defended mainly by people who have made this distinction. For most of them well-defined concepts are not the outcome of a "natural psychological production" but rather an ideal that one should respect at least in logic and science.

3.3. Most classical theories tend to regard the existence of clear-cut boundaries rather than a list of defining features as the basic characteristic of concepts

It is not so easy to accept this as a matter of fact. The problem is that most proponents of the classical theory of concepts have claimed many times and in different words, that concepts must be defined, that conceptual definitions are very important for science, and so on. Hempel, for example, holds that the definition seems to be

> the most obvious, and perhaps the only adequate, method of characterizing a scientific concept. (Hempel, 85).

It is worth mentioning here that for Hempel, as well as for most of those who insist on the importance of definitions, "a definition" does not necessary mean a list of defining features[7]. Rather more interesting is to stress that at least some of the classical authors, when describing what "the perfect concept" should be, are in reality pointing to its determinacy rather than its definition (see Sigwart, Schlick).

However, even this might not be considered convincing enough. The crucial evidence supporting the claim that the existence of clear cut-boundaries (or determinacy) of a concept is regarded by the classical theorists as the concept's most important feature, and that the definition is regarded by them basically as a tool for achieving determinacy, could be obtained when one goes more deeply into the question "why does science need well-defined concepts". When one is looking for an answer to this question, it becomes clear that definitions of concepts are not a value in themselves, they are simply a means of achieving something which is of greater importance for science: "valid judgments", "rigor of proof", "exact designation of objects". Hempel, for example, says that definitions are offered for two main purposes: to state the accepted meaning of a term that is already in use and to assign a special (new) meaning of a given term, in both cases the result is a fixing of meaning, in other words, a fixing of the boundary of the given term.

It is clear that the ultimate value for the classical theorists is the fixed meaning (the clear-cut boundaries) of a given concept and that definitions are respected as long as they are supposed to be the most appropriate, or even the only available tool for achieving that value.

The choice of the standard story, however, is not arbitrary, it follows from its assumption that the classical theory of concepts is about how concepts are represented. The main statement of a theory about conceptual representations should refer somehow to the representational notion. The defining features lists could be regarded as a mode of representation, therefore the claim that concepts are represented by means of such lists could be a central statement of a theory of concepts. Conceptual boundaries, however, could hardly be accepted as a form of representation, at least it is unclear what kind of representation it would be at all.

In other words, only those who are not set on the view that every theory of concepts should be about how concepts are represented are able to realize that the need of clear-cut boundaries of a given category or a concept is more basic for the classical view than the need of a list of defining features.

4. THE CRITICISM AGAINST THE CLASSICAL VIEW AND ITS IMPLICATIONS

Good summaries of the main arguments against the classical view are presented by Smith and Medin, Eysenck and Keene, and Lawrence and Margolis. Here, the five most popular arguments will be discussed.

4.1. The existence of disjunctive concepts

One of the basic assumptions of the classical view, according to the standard story, is that a concept is defined by means of a list of individually necessary and jointly sufficient features, which should be possessed by all instances of the given concept. In other words, the definition of a concept is nothing but a *conjunction* of defining features. There are, however, some concepts, that are defined as a *disjunction* rather than *conjunctions* of defining features. Smith and Medin give as an example the concept of a strike in baseball, where a strike could be "either a *called* or a *swinging* strike" (see Smith and Medin, 27).

The biblical concepts of clean and unclean animals, which Medin and Murphy point out as an example supporting their theory-based view, could also be regarded as disjunctive concepts, since clean animals, for example, are either creatures of the water, which have fins and scales, and swim, or creatures of the land, which have four legs.

There are also concepts that could not be represented either as a pure conjunction of defining features, or as a pure disjunction of such features. The diagnoses of some diseases often takes exactly this form. According to the *Diagnostic and Statistical Manual of Mental Disorders*

> a diagnosis of depression can be made if a disphoric mood and any five of a set of nine symptoms are present nearly every day for a period of at least two weeks. (see Medin, 96).

Smith and Medin do not consider the existence of disjunctive concepts a serious threat to the classical view but only for the reason that

> we have no firm evidence, intuitive or otherwise, about the prevalence of disjunctive concepts. (Smith and Medin, 29)

In my opinion, however, disjunctive concepts are not so rare that one can simply disregard them. The only way to preserve the classical view from the disjunctive counter-examples is to re-assess the standard story claim that having a conjunctive list of defining features is the central assumption of the classical view. If one assumes that from the standpoint of the classical view the possession of clear-cut boundaries is more important for a concept than having defining features lists, one could really get rid of the argument of disjunctive concepts, as disjunctive definitions just as the conjunctive ones determine in a clear-cut way what should be considered a member of a particular category and/or concept.

4.2. The existence of unclear cases

This argument, together with the next one, are maybe the most popular among the critics of the classical view.

Generally, one speaks about "unclear cases" when the question "Is X a member of the category (an instance of the concept) Y?" receives different answers from different persons, in other words, when different people categorize in different ways a particular object X. Of course, one could assume that the different answers, or the different solutions of the categorization problem are due to the fact that some of the people who have been asked do not, in fact, have enough knowledge about X and/or Y (see Smith and Medin, 29). In principle, if one knows the defining features list (or the defining rule) of the concept or category Y, and if one knows, or has the possibility to check, whether X possesses all defining features of Y (or satisfies the defining rule of Y), one should be able to solve the categorization problem.

Unclear cases become a real problem when a particular object X suddenly turns out to possess the defining features (or to satisfy the defining rules) of more than one (for example two) categories or concepts. Some of the greatest discussions in the history of science have been provoked by the recognition of such "unclear cases". For example, the 18th century controversy between particle and wave theories of light, the later problem of particle-wave dualism of quantum objects, the biological question as to whether viruses should be considered living things. However, there are not enough reasons to hold that the existence of unclear cases is a problem of only the classical theory of concepts. The appearance of unclear cases does not follow directly from the classical claim that concepts should have clear cut boundaries, rather it is a result of an additional methodological assumption, which does not belong to the classical theory of concepts. I mean the assumption that an object should not be a member of more than one category at the same level of abstraction: a fruit could be either an apple, or a pea, or a plum, or something else, but it can not be both an apple and a plum, or a pea and a peach.

We know, however, that due to certain genetic selection procedures some new varieties of fruits have been invented that play the role of "unclear cases" in respect of the traditional categorization. Has anybody concluded from this, that there is something wrong with our classical theory of categorization, or with our old categorization of fruits? No, the way the problem of these "unclear cases" has been solved is very simple: new categories have been introduced in order to take in the alleged "unclear" or "obscure" cases. The introduction of new categories is not the only way used to deal with the "unclear cases".

In philosophy of science Imre Lakatos has paid special attention to the different procedures that scientists could use, and actually have used in order to cope with the so-called "monsters"[8]. It is important to stress that the existence of "monsters" in science has never been regarded as a reason to give up the ideal of well-defined concepts. Most critics of the classical view, however, insist that at least in some cases we face an impossibility to draw clear-cut boundaries between different categories and to construct well-defined concepts. Insofar as this is the main, and maybe the most serious, argument against the classical view, let us discuss it in more detail.

4.3. Failure to define some concepts

First, let us put aside those examples of "failures" to define certain concepts, which either presuppose that the only way to define a concept is by means of a list of defining features, or which by "concepts" means more or less vague ideas, which, as it has been shown in the previous part of the paper, are not conceived as concepts by most classical theories. I think that Wittgenstein's famous example of the concept of "game" belongs to this group of examples. Why should we accept that we have only one concept of game and, therefore, all so-called games should possess, according to the classical view, some common features?

The fact that we use one and the same word to denote several things does not necessarily mean that these things are instances of one and the same concept. At least the classical view does not make such an assumption[9]. On the other hand, if one insists that the word game should denote one concept, one could quite successfully create a disjunctive definition of that concept: something should be considered a game either if X_1, or X_2, or X_3, or.... X_n, where $X_1 - X_n$ are the definitions of particular games. By the way, games constitute a good example of the fact that a concept could be well-defined not only by a list of defining features, but also by a set of rules, as most games are clearly defined in exactly this way – as sets of rules, rather than lists of defining features.

Second, it should be clear that the demonstration of the lack of definition of a given concept could not be considered proof that this concept is in principle undefinable. The only conclusion one could draw from the present impossibility to define a given concept is that so far nobody has succeeded in doing so.

However, nobody could infer that a successful definition will not be given to that concept in the future. It is really strange that persons who are completely aware of this (see Smith and Medin, 30) still tend to generalize the current problems which some sciences face (biology is their favorite example) to reach the strong conclusion that the classical view has a problem not only as a psychological theory but also as a "metatheoretical prescription".

The fact that a particular well-defined system of categories (concepts) meets certain problems (the most serious problems are the so-called "unclear cases") is not sufficient reason to refute the very principle of categorization. One could choose among a variety of tools to deal with this kind of problem (some of them have been already mentioned above – such as the introduction of new categories and the revision of the definitions of the old categories), while at least for the time being, and at least in science, there is no serious alternative to the classical view, regarded as a "metatheoretical prescription".

4.4. Typicality effects

Typicality effects has been traditionally regarded by the critics of the classical view as the main empirical finding giving evidence against that view. The line of reasoning of those who support this argument is the following.

The classical view claims that membership of a given category is completely determined by a list of individually necessary and jointly sufficient defining features. From this point of view all members of a given category should be equally representative, in other words, one should not expect some members of a given category to be conceived as more typical, or salient than others.

A series of psychological experiments, however, have disproved this expectation, showing that people tend to ascribe typicality ratings to the members of *all* categories (to the instances of *all* concepts), even to a member of well-defined categories such as that of even numbers[10]. At the same time it has been shown that typicality ratings of the members of a given category do not correlate very well with the frequency or familiarity of the members (instances) (Mervis, Catlin, and Rosch). One could conclude from that (and many do concluded) that these ratings are produced by (are a function of) some internal structure, which definitely *is not* a list of defining features and as such, could not imply any typicality ratings.

The only correct conclusion, however, that one could draw from the analysis of the results concerning the typicality effects is that definitions have nothing to do with typicality. All other conclusions are based on some additional assumptions, most of them very problematic as we tried to show in the previous part of this paper. The least tenable conclusion that could be made at this point is that the typicality effects violate the classical view on concepts and categories.

If one identifies the classical view only with the "methodological prescription" that categories should have clear-cut boundaries, and therefore concepts should be defined (not necessarily by means of lists of defining features) in a way that ensures clear-cut boundaries between instances and non-instances of those concepts, one has no reason to see any danger in typicality effects. The only function of the definition of a concept is to draw a boundary between members and non-members of the category this concept is referred to. The definition has nothing to do either with the *mental representation* of that category, as the standard story wrongly assumes, or with the way the alleged mental representation *is processed* in the mind. In particular, the definition of a concept has nothing to do with the answer to the question as to which instances of that concept one considers more typical. The experimental results reveal that typicality effects do not say anything more. Even the conclusion that the typicality rating is a function of the conceptual mental representation is not well grounded. How could one be sure that in the judgment whether apples or peaches are more typical fruits, the only factors that influence the decision are the structures of the mental representation of apples, peaches and fruits but not, for example, some elements of the context in which these concepts have been used?

4.5. Use of non-necessary features to represent categories

Some evidence of the use of nonnecessary features (for example, "flies" for a bird) in the categorization task is found in the experimental results of Rosch and Mervis (1975), Hampton (1979), see also (Smith and Medin). At first glance, this argument seems to be serious, as it concerns the process of categorization and the only function of the definition of a concept, according to the revised version of the classical view (as defended here), is exactly to support categorization, i.e., to distinguish instances from noninstances of a given concept. The fact that sometimes people rely on nonnecessary features when deciding whether a particular object is an instance of a given category may have different interpretations and none of them is crucially incompatible with the classical view. First, the definition that one uses might be imperfect, i.e., including nonnecessary features; the classical view allows this. Second, most classical theories insist on the difference between the concept represented by its definition and the mental representation of the category this concept refers to. Some nonnecessary features could be included in this mental representation, and this mental representation could also be used as a tool for solving the categorization task. Conceptual definitions support the precise categorization. In everyday life, however, even if one has at one's disposal the precise definition delineating a particular category, one could find it much more convenient to use the imprecise mental representation of the same category, which is either more "at hand", or simply works faster. This hypothesis also does not contradict the classical view. Therefore, without any regard to the interpretation one gives to the use of nonnecessary features in the categorization task, one could quite well bring these experimental findings into line with the classical view on concepts and categorization.

5. CONCLUDING REMARKS

In the 1970s, the experimental findings of Eleanor Rosch drew attention to the fact that people categorize the perceived world not only by means of the construction of definitions but also by the formation of prototypes. These findings have been broadly recognized as the beginning of the so-called categorization debate. The real debate, however, was opened some years later, when most of the cognitive science community reached an agreement that conceptual definitions have nothing to do with the psychological process of categorization and that if such definitions (could) exist at all, they play a small role, if any, in real cognitive processes, and therefore, the view whose main claim is that concepts and categories are represented by means of sets of defining features (the so-called classical view) is to be recognized as inadequate and replaced by a more sound theory. Some of the critics of the classical view have even claimed that this view has failed not only as a psychological theory but also as a methodological prescription in natural sciences (especially in biology).

This paper is an attempt to clarify that the notion of the classical view accepted by the standard story of the categorization debate is based on three wrong assumptions: (1) that the classical view is about what concepts are (how they are represented), (2) that from the classical point of view concepts are mental representations, and (3) that all classical theories assume that concepts are represented by means of lists of defining features. In fact the most influential classical theories of concepts are normative or prescriptive theories, i.e., they are about what a perfect concept should be; most classical theories make a sharp distinction between ideas (images), which are mental representations and natural psychological productions on the one hand, and concepts, which might not have a mental reality and which are created on purpose on the other hand; and in the end, for most classical theories the most important feature of a concept (a category) is its clear-cut boundary, the definition is regarded only as a tool to achieve a sharp distinction between members and nonmembers of a given category (instances and noninstances of a given concept). What follows is that the theory-based view on concepts and categorization that in the 1980s was introduced as an alternative to the classical view and should not be regarded as an alternative but rather as a variant (or generalization) of the classical view insofar as it also conforms to the clear-cut boundaries of categories and concepts.

Most of the criticism against the classical view would only be true if the three assumptions listed above were true too, but as this is evidently not the case, the main critical arguments against the classical view tend to be untenable. If one accepts that the conjunctive lists of defining features are not the only tool to form a well-defined concept, it becomes evident that the so-called disjunctive concepts are not counterexamples for the classical view and that most of the alleged failures to define certain concepts are only failures to define them by means of lists of defining features. In principle, the clear-cut boundaries of concepts and categories do not leave room for unclear cases. When such cases appear, they are due either to lack of knowledge, or to the inappropriateness of the particular system of categories. Typicality effects and the use of nonnecessary features in the categorization process become a problem for the classical view only if one assumes that the definition of a given concept is at the same time the only mental model of that concept one uses when solving real categorization tasks. In the end, there is no serious evidence against the merits of the classical view as a normative (regulative or prescriptive) principle of concept formation in science. The problems that particular scientific theories meet (unclear cases, difficulties in defining certain concepts) have always been solved not by escaping from the classical principle of categorization but rather by improvements and reorganizations of the particular categorical systems adopted by those theories.

NOTES

[1] Wittgenstein (1953) gave as an example the concept of "game". It is impossible, holds Wittgenstein, to find out a feature that is common for all games, since some games, for example, are competitive but others are not, some games need more than one player but there are also games which could be played individually, some games presuppose a winner but the child's game of ring-around-a-rosy has nothing to do with winning, some games are played for pleasure, but definitely not all of them, and so on.

2 Some authors call it the *natural* view (Gardner) or the *probabilistic* view (Smith and Medin).

3 This view is also known as *theory-theory* view (Margolis and Laurence) and *explanation-based* view (Eysenck and Keane).

4 The only exception that I know is Smith and Medin's note that "the classical view has also served as a metatheoretical prescription of what scientific concepts should look like" (Smith and Medin, 31) and that even as a metatheoretical prescription this view runs into serious problems. More attention to the arguments of Smith and Medin will be paid in part 4 of this paper.

5 Following Aristotle, most philosophers call "categories" some special sets of basic concepts. In psychology categories are viewed either as another name for concepts (for all concepts) or as groups of things that the concepts stand for.

6 Smith and Medin notice that this view of Aristotle "has always had its critics" (Smith and Medin, 2) but only in the past decade has this criticism become more intense, giving rise to new views.

7 Bridgman, for example, introduces the notion of the "operational definition" – a definition that specifies the meaning of a scientific term by means of a testing operation providing the criterion for the application of that term (see Hempel, 88).

8 This is Lakatos's special term for what psychologists today identify as unclear or non-typical instances of a concept.

9 Indeed, there is a broadly-accepted definition of a concept in linguistic circles according to whom concepts are the meanings of the words, but even this definition (which is controversial enough because we also recognize as concepts entities that are named by more than one word – such as a pet fish or a black hole, for example – or even by a whole expression – as deep-sea diver, or proper decimal fraction, or least common multiple), does not imply that one word could not have more than one meaning, therefore, denote more than one concept.

10 Rips, Shoben and Smith (1973), and Rosch (1973) are maybe the first to have measured the typicality ratings of the members of natural categories such as mammals, birds, and fruits. It has been shown, for example, that robin and sparrow are considered as typical birds, and chicken and penguin atypical. The fact that even well-defined concepts such as the concept of even number demonstrate typicality effects was established by Armstrong, Gleitman and Gleitman (1983).

REFERENCES

Armstrong, Sharon L., Lila R. Gleitman, and Henry Gleitman. "What Some Concepts Might Not Be." *Cognition* 13 (1983): 263-308.

Brown, Roger, and Eric H. Lenneberg. "A Study in Language and Cognition." *Journal of Abnormal and Social Psychology* 44 (1954): 454-462.

Eysenck, Michael W., and Margaret Keane. *Cognitive Psychology*. Hove: Lawrence Erlbaum, 1995.

Fodor, Jerry A. *Concepts. Where Cognitive Science Went Wrong*. Oxford: Clarendon Press, 1998.

Frege, Gottlob. *The Foundations of Arithmetic*. 2nd ed. Oxford: Basil Blackwell, 1986.

Gardner, Howard. *The Mind's New Science. A History of the Cognitive Revolution*. New York: Basic Books, 1985.

Hampton, James A. "Polymorphous Concepts in Semantic Memory." *Journal of Verbal Learning and Verbal Behavior* 18 (1979): 441-461.

Hempel, Carl. *Philosophy of Natural Science*. Englewood Cliffs, N.J.: Prentice Hall, 1966.

Lakatos, Imre. *Proofs and Refutations: The Logic of Mathematical Discovery*. Cambridge: Cambridge University Press, 1976.

Margolis, Eric. "A Reassessment of the Shift from the Classical Theory of Concepts to Prototype Theory." *Cognition* 51 (1994): 73-79.

Margolis, Eric, and Stephen Laurence. (Eds.) *Concepts. Core Readings*. Cambridge, MA: The MIT Press, 1999.

Medin, Douglas L. "Concepts and Conceptual Structure." In: Thagard, Paul. (Ed.) *Mind Readings*. Cambridge, MA: The MIT Press, 1998: 93-125.

Mervis, Carolyn B., J. Catlin, and Eleanor Rosch. "Relationships Among Goodness-of-example, Category Norms, and Word Frequency." *Bulletin of the Psychonomic Society* 7 (1976): 283-284.

Murphy, Gregory L., and Douglas L. Medin. "The role of theories in conceptual coherence. "*Psychological Review* 92 (1985): 289-316.
Rips, Lance J., Edward J. Shoben, and Edward E. Smith. "Semantic Distance and the Verification of Semantic Relations." *Journal of Verbal Learning and Verbal Behavior* 12 (1973): 1-20.
Rosch, Eleanor. "On the Internal Structure of Perceptual and Semantic Categories." In: Moore, T. E. (Ed.) *Cognitive Development and the Acquisition of Language.* New York: Academic Press, 1973: 111-144.
Rosch, Eleanor. "Principles of Categorization." In: Margolis, Eric, and Stephen Laurence. (Eds.) *Concepts. Core Readings.* Cambridge, MA: The MIT Press, 1999: 189-206.
Rosch, Eleanor, and Carolyn B. Mervis. "Family Resemblance Studies in the Internal Structure of Categories." *Cognitive Psychology* 7 (1975): 573-605.
Schlick, Moritz. *General Theory of Knowledge.* New York: Springer-Verlag, 1974.
Sigwart, Christoph. *Logic.* Vol.1. New York: Macmillan & Co, 1895.
Smith, Edwart E., and Douglas L. Medin. *Categories and Concepts.* Cambridge, MA: Harvard University Press, 1981.
Wittgenstein, Ludwig. *Philosophical Investigations.* New York: Macmillan & Co, 1953.

SLAVIAN RADEV

THREE WORDS: HYPERTEXT AND ARGUMENTATION READINGS OF TRACTATUS LOGICO-PHILOSOPHICUS

This paper is a comment to the hypertext and argumentation system readings (http://aragorn.pb.bialystok.pl/~radev/tlf/tlf.htm, or http://members.spree.com/college/slavian/tlp/tlf.htm) of the Wittgenstein's Tractatus Logico-Philosophicus. It is a speculation that the author may use such a form of presentation of his ideas if he has the corresponding instruments.

> Some Frenchman – possibly Montaigne – says: "People talk about thinking, but for my part I never think, except when I sit down to write." It is thus never thinking, unless when we sit down to write, which is the cause of so much indifferent composition. But perhaps there is something more involved in the Frenchman's observation than meets the eye. It is certain that the mere act of inditing, tends, in a great degree, to the logicalization of thought.
> Edgar Allan Poe, *Thoughts and words,* (Poe, 312)

1. INTRODUCTION

The inspiration of that investigation is based on the idiotic question "What did the author want to say?" and on the corresponding answer "If the author wants to say, why did he do not say?". A consequence of these questions is to prepare text in a dialogical form that includes all possible readings that the author allows. Tractatus Logico-Philosophicus (TLP) is written in a very special form that allows many reading sequences. I think that way is shown in the WWW pages above, where one may read the text of the TLP following the author instructions. Arguments in favour of that are the footnote and the preface.

2. HYPERTEXTS AND ARGUMENTATION

2.1

Hypertext is the best known form of presentation in Internet WWW pages. The HTML language possibilities allow us to include a reading algorithm in the document (WWW page) together with all "cookies".

2.2

Argumentation system (Radev 1996) is a quadruple <U,A,S,v> where:

Dimitri Ginev (ed.), Bulgarian Studies in the Philosophy of Science, 163-166.

U = {p,q,r,...} is the set of propositions;

A = {a,b,...} is the set of argumentations (binary relations between propositions); <p,q>∈a means that p is an argument for q in a-sense;

V:U→{0, 1} is a valuation of the propositions from U in the set {0, 1} (or more precisely in the set of values: "accept", "reject", "unknown" denoted 1,0,? respectively).

S={s,t,...} is a set of argumentation strategies. Argumentation strategy is an algorithm to calculate the value of the thesis from the values of its arguments.
In this paper we use only the following strategy: accept if all the arguments are accepted (evaluated 1), reject if some of the arguments are rejected (0), try to evaluate (looking for its arguments) if the value of the argument is unknown (?).

3. READING POSTULATES

3.1

If I accept the thesis then I need no arguments for it.

3.2.

If I agree with the arguments and agree that these arguments are sufficient, then I agree the thesis.

1-2. Hence, after I read the argumentation and accept the thesis, then I no longer need the arguments used in the proof (see also TLP 6.54).

3.3.

If I use logical consequences it must be used successfully.

3.3.1.
The most common logic is based on a simplified "modus ponens" rule (the transitivity of the relation "is"):

from "A is B" and "B is C" (logically) follows "A is C"

3.4.

I do not know everything. Every new piece of information changes my knowledge (system).

4. STRUCTURE OF TLP

maximal length 7 (including the close brackets)
maximal width 7 (including 0-labeled comments)
maximal depth 6 (including the main numbers f.i. 5.47321)

maximal inner width 7 (of the subtrees) except 3 sentences, one of which is 4.128 that says "The logical forms are *anumerical*"

Hence: with small cosmetic changes it is in a 6x6x6 cube (without notes) and in 7x7x7 cube (almost the full text).

It is speculation whether the book says something with that dimension, but it is clear that the complete text is observable (at once) in any direction.

5. MODALITIES

There are a lot of modalities in TLP independently of the fact that it is written in three words. In those readings I do not consider these modalities because the title (also in three words) is a sufficient modality. Almost all points of view are presented in the arguments (comments) and if they are coloured then the outcome colour is white.

I use in the WWW page only one modality – the colour red for the sentences used in the actual logical step. This is a mistake. It is out of the convention to underline that which has to be forgotten. I underline in such a way only the arguments that are used in those readings only.

In the proposition "From A follows C" we forgot the premises "From A follows B" and "From B follows C".

6. THE READINGS

The proposed readings in the pages (http://aragorn.pb.bialystok.pl/~ra-dev/tlf/tlf.htm, http://members.spree.com/college/slavian/tlp/tlf.htm) are based on the following observations:

6.1

TLP is within the context of the title (in three words), the motto and the last sentence. The first two say to look for three words, the last (and the last but one) say that these words are written. To define the context I use the brackets "[" and "]".

6.2.

TLP is written in a tree-like structure (see footnote about the meaning of the structure). The form of the structure is: root (invisible), 6 main points, their arguments/comments...

6.2.1

The root is not written because it follows logically from the main points. It has to be deduced.

6.2.2

The main operations are logical: the argument for that is the order of the words in the title.

6.3.

The comments differ from the main points that are used in the logical consequence, not in the argumentation/explanation text. From the comments we do not deduce anything.

7. NOTE

For me the current paper is written in √3 words. Sorry that it contains 1018 words.

motto:

> ... gimme a plate of ham and eggs – how much? – and – do you love me, kid?
> Carl Sandburg, *Threes* (Sandburg, 203)

REFERENCES

Poe, E. A., *Poems and essays.* London: J.M. Dent & Sons Ltd., 1926.
Radev, S. "Argumentation systems," *Fundamental Informaticae* 28 (1966): 331-346.
Sandburg, C. *Selected poems.* New York: Harcourt, 1926.
Wittgenstein, L. *Tractatus Logico-Philosophiciis.* London: Routledge, 1996.

Part V

PHILOSOPHY OF SCIENCE
AND THE CONTINENTAL IDEAS

ANGUEL S. STEFANOV

ON KANT'S CONCEPTION OF SPACE AND TIME

1. BRIEF INTRODUCTION

This essay is an impartial attempt at fencing with Kant's transcendental aesthetical doctrine of space and time against some well known types of criticism. By saying "impartial", I mean only that one of my aims here will be a proper estimation of the relevance of the critical arguments involved, and not a purposeful commitment with some philosophical viewpoint, be it Kantian, or not.

In the second section I show that the criticism of Kant's transcendental conception of space and time could be differentiated into *two main types*. While the first type of criticism is a standard reaction of a general character, defining inconsiderately Kant's conception to be a piece of subjectivism, the second type finds it to be an obsolete attempt at perpetuating the universality of 3-dimensional flat space. The first type of criticism is engendered by a misleading change of concepts – the tacit laying of realist notions into Kantian-like expressions. The second type of criticism is based on the contemporary development of natural science and mathematics.

The third section presents R. Carnap's and H. Reichenbach's conviction that a clear-stated difference should be made between physical geometry and mathematical geometry. In so far as Kant did not make it, he declared in an aprioristic way the objective validity of Euclidean geometry, and short-sightedly precluded the possibility that physical space could have a non-Euclidean character – a fact that is now accepted by relativistic physics, not to speak of contemporary super-string theories, situating the universe in multi-dimensional (e.g., 10-dimensional) space-time. My aim is not to criticize Carnap's and Reichenbach's view of space and time, but to show that *their critique of Kant's transcendental aesthetic is merely irrelevant*. In doing so I argue that the physical/mathematical distinction of geometry is not a Kantian problem.

In the last section *a rather* "exotic" thesis is put forward. It considers *the possibility for introducing a concept of space as a theoretical construct in Kant's philosophy*, as if there is a physical medium of dynamic phenomena. In this sense it plays the rôle of an *auxiliary concept*, leaving Kant's transcendental approach (regarding space and time as pure forms of human intuition) untouched. In fact, Kant himself admitted the hypothetical existence of spatial regions with properties and dimensions different from 3-dimensional flat space in a pre-critical work – a fact that seems to be not so familiar. However, orthodox kantians might be inclined

Dimitri Ginev (ed.), Bulgarian Studies in the Philosophy of Science, 169-185.

to reject the suggested conceptual complication. One ought then to admit the *incommensurability* of Kant's transcendental conception and the general realist view – an alternative that at least protecs Kant's philosophy from direct criticism.

2. TYPES OF CRITICISM AGAINST KANT'S TRANSCENDENTAL CONCEPTION

I group different criticisms against Kant's transcendental conception of space and time into two main types.

The first type of criticism represents a standard reaction of a general character. Kant's conception of space and time – transcendental aesthetic – is taken as a whole to be some kind of subjectivism, and often – in the tradition of eastern dialectical materialism – of subjective idealism. It is natural then, that having been led by this qualification, realist philosophers would reject Kant's conception from the very outset, as being incompatible with the principles of realist epistemology.

> Kant's teaching about space and time is subjective and idealistical

an author of a voluminous study of the philosophy of the Königsberg thinker clearly declares (Karapetyan, 160). And even more:

> The real basis of Kant's phenomenalism, and of his subjective idealism in general, is the teaching about the ideality or subjectiveness of time and space. All efforts of Kant to draw a line between himself and the resolute idealism of Berkeley are unsuccessful, since he builds his transcendental idealism namely upon the teaching about the ideality of time and space. (*Ibid.*, 155-6)

Once Kant's conception of space and time is taken to contain the same metaphysical blend as that of Berkeley's philosophy, and to proclaim in a subjective-idealist manner the ideal character of space and time, it becomes quite understandable why it should be criticized in principle by every staunch realist.

I named this type of criticism "a standard reaction of a general character". The criticism is standard as a motivation – a philosophical position is being declined because it is in contradiction to another one. At the same time the criticism has a general character, since it does not appear as a result of a consistent grasping of Kant's conceptual system, and is based upon a general, but misleading, premise. This is the premise that Kant's transcendental conception of space and time is (subjective-) idealistical.

It is well known that Kant does not think of himself as being an idealist, he does not share the philosophy of Berkeley, and calls this philosophy "dogmatic idealism". In the case considered, however, the common prejudice that Kant's conception of space and time is a piece of subjective idealism is created by an unintentional change of concept.

This change is expressed in the fact that materialists who give an assessment to transcendental idealism lean on their own view of space and time. They lay *their* notion in the Kantian statement that space and time refer to the knowing subject, but not to real things, which objectively exist independently of him. *The realist concept of space and time has a quite different meaning from that of the Kantian aesthetical concept*; and if we have actually *set as a precondition* the realist (materialist) concept in a formally Kantian utterance, then we would have unavoidably stated a subjective-idealistic thesis.

Kantian space and time are found to be "with" the perceiving subject, and not to be noumenal essences or objective relations, because, according to him, they compose the conditions for the possibility of human experience. If Kant would have contended that space and time taken as external givenness for the subject, as *objective* relations among directly knowable things, are *subjective*, then he would have stated not a sheer subjective-idealistical thesis, but rather a conspicuous contradiction.

Kant's concepts of space and time will be examined in the next section. However, even now it becomes clear that one may not "blame" Kant for being a subjective idealist, which he really is not. The only criticism, which then might be addressed towards Kant is, why he puts just his transcendental meaning in the concepts under consideration, and evades their realist meaning. All the same this meaning had not been something foreign or strange to him during his so called pre-critical creative period. However, this (otherwise quite interesting) question can play the rôle of a critical argument, just as much as the belief that making philosophy must follow ready-made conceptual standards, can serve as a revelation.

The second type of criticism against Kant's transcendental aesthetic is also rooted in a misleading premise, which is rarely explicitly formulated. However, it exhibits the feature of a telling criticism, since it explores the inherent geometrical characteristics of the proposed space, and is based upon the post-Kantian development of mathematics, physics, and cosmology. What is this type of criticism about?

Kant puts forward an original *philosophical* conception of space and time, or *the* space and *the* time. By inserting a definite article in front of the last nouns I would like to point to their one-ness, or uniqueness, having also in mind that the pure intuitions expressed by them are referent to the sphere of the knowing subject; and there one can hardly imagine, even less postulate, the presence of *different* spaces. The Königsberg thinker brings out the gnoseological rôle and functions of space and time, and not so much their mathematical characteristics. There is no doubt that, if we ought to involve the accepted terminology, then Kantian space should be determined as being flat (having no curvature), or three-dimensional Euclidean space. Time is one-dimensional and continuous. The objections of too many authors (philosophers and scientists), roughly speaking, since the middle of the nineteenth century onwards, are based on the fact of the collection of a nosegay of non-Euclidean geometries. Some of them – such authors argue – could represent the qualities of physical space and time in a better way than is within the possibility of Euclidean geometry, which is unproblematically applied to events taken only from the classical Newtonian world. Even a sparing critical conclusion, which is alluded to, is that Kant's commitment to only Euclidean geometry is a manifestation of a methodological improvidence.

Pavel Florensky, a connoisseur in the thematical field of space and time both in positive science and fine arts, wrote in 1924, that

> one hundred years ago N. I. Lobachevski stated the resolute anti-Kantian thought, which remained then only as a bold dictum, namely that different phenomena of the physical world take place in different spaces, and are, therefore, subordinated to the corresponding laws of those spaces. (Florensky, 5)

The presence of "different spaces" is at variance with the Kantian assumption of only one space. Florensky himself maintains the view that geometry is an empirical science. The geometry of space is determined, according to him, by the field of forces and the latter, vice versa, by geometry.[1]

Recording rightly that Kantian time and space are neither self-subsistent entities, nor empirical generalizations, but *a priori* forms of sensory perception, being necessary conditions of every experience, Milič Čapek still insists that

> the Newtonian model of the universe with both its components – Euclidian space and mathematically continuous time – is beyond the danger of being challenged by *any* further experience. No wonder that quite a number of *neo-Kantians were not happy when this bold prediction of their master clashed with the new trends in physics and geometry.* [2]

I will try to show later how the last typical criticism of Kant's transcendental conception of space and time becomes possible. Even here I would like to clarify a methodological premise, which is taken for granted by Kant's critics, and thus is tacitly involved in their argumentation. However, the premise is neither evident, nor ensued from Kant's transcendental viewpoint. However, it is quite simple.

It is a necessary (and not sufficient) condition for a conception "to clash" with another one, that they possess the same cognitive status. So, *the premise is* that Kant's conception of "Euclidean space and mathematically continuous time" bears the same cognitive status, or is of the same epistemological character, as the one implied by "the new trends in physics and geometry", i.e., as the conception that the geometry of physical space-time is *not* Euclidean. We shall see, however, that this is not the case. Kant's conception, as an integrative part of his transcendental epistempology, has a quite different rôle and pretention than those of contemporary *theoretical models* of space-time manifolds. The cognitive pretentions are rather philosophically incommensurable, than clashing with each other, so I cannot definitely say whether neo-Kantians should, or should not be happy with the new trends of physics and geometry. Yet I am inclined to think that Kant himself would have no reason to worry about that.

There are even more trenchant assessments of Kant's conception to the effect that it is merely false. Such "blunders", Zucker contends for instance, as are his *a priori* arguments in support of Euclidean geometry, absolute space and Newtonian mechanics, "spell the collapse of Kant's entire attempt". (Zucker, 480) This type of criticism is shared, though to different extents, by many philosophers and scientists, expressing different attitudes towards Kant's ideas. In this whole spectrum of related opinions I shall be mostly interested in the argumentation of such thinkers like *Rudolf Carnap* and *Hans Reichenbach*, who have proved for a long time their methodological competence and creative capacity. All the more that their critical attitude towards Kant's conception of space and time has been espoused by many first-class philosophers, and has become paradigmatic, or at least very popular, within contemporary philosophy of science. It suffices to mention Adolf Grünbaum's famous book *Philosophical Problems of Space and Time*, in which the reader is notified – as something well established – that

> Reichenbach made a particularly telling contribution to the disintegration of the Kantian *metrical* a priori of *visual* space by showing that such intuitive compulsion as inheres in the Euclideanism of that space derives from facts of logic in which the Kantian interpretation cannot find a last refuge...[3]

My interest in the criticism they launch is also motivated by the fact that the conceptions of these philosophers are entirely acceptable by, or at least completely comprehensible for everyone who is tending towards philosophical realism. This means here that when displaying my dissention with the statements of R. Carnap and H. Reichenbach, I *am not* going to criticize *their own* conceptions of space and time, and their view about the scientific status of non-Euclidean geometries, but I shall make an attempt to show their misunderstanding of Kant's transcendental viewpoint. I find the importance of such an attempt to be twofold. In the first place it would lead to the merited vindication of a "first value" philosopher (in relation to the alleged falsity of his conception, visible on the grounds of today's achievements in mathematics and physics); and in the second place the attempt, if successful, could initiate the thesis about the simultaneous validity of Kant's transcendental setting and the broadly shared concept of space in realist philosophy of science (see section 4).

3. HAS KANT MADE AN ERROR?

Having been guided by his respect towards the Königsberg thinker, Rudolf Carnap even makes an endeavour to whitewash the otherwise criticized Kant's conception:

> *Kant should not be blamed for his error* because, in his day, non-Euclidean geometry had not been discovered. *It was not possible for him to think about geometry in any other way.* In fact, throughout the entire nineteenth century, except for a few bold individuals, such as Gauss, Riemann, and Helmholtz, *even mathematicians took this Kantian point of view for granted.*[4]

But has Kant made an error?

For a reader not tempted by Kant's philosophy, after acquaintance with Reichenbach's and Carnap's reasoning, the answer may well be positive.

Let us see now what the error, imposed on Kant, consists of? One may easily conclude, even from what has been said up to now, that the error consists in postulating the uniqueness and originality of the directly perceived features of space and time, and in not admitting the possibility of even essential corrections, coming from future theoretical and/or empirical scientific discoveries. Kant was a child of his Newtonean epoch, and (even he) could not rise over its conceptual framework. But the fruitful development of mathematical investigations during the 19th, and especially during the 20th century, have opened the possibility for the construction – in a non-contradictory way – of various mathematical spaces, based on non-Euclidean axiomatics. The axioms, or the initial statements of formal constructions, *cease to be looked upon as obvious truths any longer*, as the recognized tradition has required for ages, from ancient times to the middle of the 19th century. H. Reichenbach clearly explains that

> for geometry as a mathematical science, there is no problem concerning the truth of the axioms. This apparently unsolvable problem turns out to be a pseudo-problem. The axioms are not true or false, but arbitrary statements. It was soon discovered that the other axioms could be treated in the same way as the axiom of the parallels... The discussion of this statement leads away from mathematics; as a question about a property of the physical world, it is a *physical* question, not a *mathematical* one. This distinction, which grew out of the discovery of non-Euclidean geometry, has a fundamental significance: it divides the problem of space into two parts; the problem of mathematical space is recognized as different from the problem of physical space. (Reichenbach, 5-6)

Carnap's specification of Kant's error goes just along this line of reasoning:

> Today, it is easy to see the source of Kant's error. It was a failure to realize that there are two essentially different kinds of geometry – one mathematical, the other physical. (Carnap, *Foundations*, 181)

Mathematical geometry is an abstract deductive system, based on axioms "that do not have to be interpreted by reference to any existing world". In this sense *it is a pure construction.* "Physical geometry, on the other hand, is concerned with the application of pure geometry to the world." [5]

Carnap goes still further in his critique of Kant's philosophy. After a presentation in a contemporary and understandable manner of the distinction drawn by Kant between analytical and synthetical judgments, he declines the answer to the central problem of Kant's transcendental approach: do synthetical *a priori* judgments exist and how they are possible? As opposed to Kant, Carnap raises the doctrine that there are no such judgments in mathematics, and one can still less imagine how they are possible in a natural science like physics.[6] Mathematical geometry, according to Carnap, refers to pure mathematics. As a pure science it is really aprioristic (i.e., no statements with an empirical content are interwoven in it), but this geometry is not synthetical, it is analytical. This is so, because mathematical geometry is a deductive system whose axioms are provided with no interpretation at all, that could refer them to some fragment of the existing world.

> Once this distinction between pure and physical geometry is understood, it becomes clear how Kant's belief, and the beliefs of almost all nineteenth-century philosophers, involved a fundamental confusion between two fields of quite different character. (Carnap, *Foundations*, 182-3)

Now, it is easy for me to reconstruct the coinciding criticisms of Reichenbach and Carnap against "Kant's belief". They rest upon two lucid theses:

(1) Two types of geometry must be clearly differentiated – mathematical, and physical.

(2) Kant does not make the distinction (1), and to this effect he does not admit the possibility some other, different from the Euclidean geometry, to be applicable (at least with the same, if not better success) to the natural world.

To all appearances the immediacy of this criticism has convinced a lot of contemporary authors to accept it as a cogent argument against Kant. "Why was Kant wrong?" – the question was explicitly raised by V. L. Hramova in her interesting book about the relation between theory and experiment in relativistic physics. The direct answer, which she finds to be the right one, is as follows: "The trouble lies in the fact that he had not differentiated the two kinds of geometry – the mathematical and the physical one" (Hramova, 173). The same critical argument stands in the presumably most authoritative university textbook in philosophy of science of our days, prepared by Martin Curd and Jan Cover. One can read there that "whether Euclidean geometry is true of space is an empirical matter, not something we can know a priori as Kant and Frege had thought", since after the success of Einstein's general theory of relativity we have to distinguish between the two types of geometry, mentioned above (Curd and Cover, 370).

I have two reasons not to express any doubts about thesis (1). The first one is the plausability of this thesis, which is acceptable from a realist point of view, especially after Carnap's introduction of the "explicates" of L- and F-true propositions, i.e., after the suggested distinction between necessary, analytical truths, and factual or synthetical truths. It is also well known in this respect, that since Hermann Minkowski ascertained the 4-dimensional pseudo-Euclidean character of physical space-time in 1908, and especially since the creation of the general theory of relativity (and the methodological discussions engendered by it), the so called problem of the *complementarity* between geometry and physics has become one of the most debated in the philosophy of science. In fact, this is a problem about the choice of a convenient mathematical structure – a geometrical space having definite metrical and topological properties – which can describe in a feasible manner the thing, called by contemporary scientists "real physical space-time".

The second – and essential – reason to leave (1) without further comment is the following. If the thesis is taken to be false, then the critical argument (2) against Kant loses its validity, and thus my task is accomplished. What I am interested in here, however, is to point out that thesis (2) is an *irrelevant* criticism of Kant *notwithstanding the fact how true (1) might be.*

Why is (2) an irrelevant criticism of Kant's transcendental conception? The lapidary answer is, because Kant (in his *Critique of Pure Reason*) is not facing the problem about the description of space and time as being outside the perceiving subject, i.e., about their description as some objective things or relations, standing against the cognizing spirit. However if this problem is beyond his philosophical transcendental setting, then one could hardly speak of any "fundamental confusion" Kant has allegedly made "between two fields of quite different character" – between pure (mathematical) and physical geometries.

Thesis (2) could still be a relevant critical argument, only if Kant had ascribed a *primary validity* to three-dimensional Euclidean geometry, which, rooted in its self-evidence, and thus, by declaring its uniqueness, had attributed to it an ontological status. Reichenbach's interpretation seems to be of this kind. According to him, the Kantian doctrine

> contends that an *innate property* of the human mind, the ability of visualization, *demands that we adhere to Euclidean geometry*. In the same way as a certain self-evidence compels us to believe the laws of arithmetic, a visual self-evidence *compels us to believe in the validity* of Euclidean geometry. It can be shown that this self-evidence is not based on logical grounds. (Reichenbach, 32, my italics)

Reichenbach substantiates the last statement by proving a theorem, that stands at the heart of the already mentioned problem of complementarity between geometry and physics, which is presented in his book as *the principle of the relativity of geometry*. The epistemological moral of this theorem[7] is, that we cannot meaningfully speak of *any* geometry of space to be the *true* geometry. This is so, because only a *combination* between geometry and a universal field of force can produce empirically verifiable claims about physical reality. Thus a freedom exists for our choice of spatio-temporal geometry, and this is stated by the principle of the relativity of geometry. On the base of this principle it is easy to see why the preference for Euclidean geometry, compelled by a visual self-evidence, is void of any logical argument.

Reichenbach's view, followed by many philosophers of science, is beyond any doubt, with the exception of the *type of apriorism* imputed to Kant. If Kant has claimed indeed, that human ability of visualization "demands that we adhere to Euclidean geometry", that some *innate property* of the mind makes this geometry self-evident to us, and thus "compels us to believe in its validity", then the conclusion of Reichenbach about the absence of logical priority of any geometry whatsoever would represent a strong critical argument against this type of apriorism. But Kant's apriorism *is not of this type*.

Here is the place probably for a curt remark, that even a prominent philosopher like Sir Karl Popper, with all his great methodological respect for Kant, shares some part of this Reichenbachean position. I find the fact to be curious, in so far as, to my knowledge, Popper has never been intellectually influenced by Reichenbach. Anyhow, he writes that Kant has made an error, although this error detracts in no way from his magnificent achievement.

> What was his error? As I have said, Kant, like almost all philosophers and epistemologists right into the twentieth century, was convinced that Newton's theory was *true* ... It was an unavoidable error – unavoidable, that is, before Einstein. (Popper, 190-191)

It is my contention that Reichenbach's and Popper's "error", to exploit the same word, lies in the presupposition that Kant's aprioristic viewpoint *objectifies* in some way, or another, the forms of intuition and the synthetical principles of the pure understanding upon nature. Going back to Reichenbach, this is expressed in the alleged *validation* of three-dimensional Euclidean geometry (because of the innate property of the human mind – the ability of visualization), which validation is but an implicit ontological interpretation. Such an interpretation, both aprioristic and ontological, is very easily criticized. As for Popper, my objection concerns his conviction that Kant allegedly claimed the possibility of imposing final laws upon nature, in the sense that he believed in their immutable truth:

> When Kant said, 'Our intellect does not draw its laws from nature but imposes its laws upon nature', he was right. But in thinking that these laws are necessarily true, or that we necessarily succeed in imposing them upon nature, he was wrong. (Popper, 48)

As I have already noticed, Kant's apriorism is not of this type. His critical standpoint, I suppose, prevents him from validating the truth of a geometry, be it Euclidean, or not, being guided by some unclear innate properties of a aprioristic nature. He neither writes explicitly, nor even alludes to our faculty of receiving representations (Vorstellungen), or of judgment, *compels us to believe in*, or demands that *we adhere to* Euclidean geometry. I think that the essence of Kant's transcendental aesthetic is not in the fact that something makes us to adhere to a definite geometry with the aim to impose it on the space outside, but that space and time by themselves are necessary *conditions* for the knowing consciousness to be able to possess any intuition (Anschauung) whatsoever. For Kant

> Space does not represent any quality of objects by themselves, or objects in their relation to one another; i.e. space does not represent any determination which is inherent in the objects themselves, and would remain, even if all subjective conditions of intuition were removed... Space is nothing but the form of all phenomena of the external senses; it is the subjective condition of our sensibility, without which no external intuition is possible for us. (Kant, *Critique*, 26; A: 26, B: 42)

Time also, according to Kant, is not something existing by itself, and cannot be an objective determination of external things, but is the form of the internal sense, of our intuition of ourselves (*ibid.*, 30-31; A: 32-33, B: 49)

All this means that under the terms of space and time Kant understands *a priori* forms of the subject of knowledge, *which make possible the very having of any sensuous intuition.* It is well known that knowledge is impossible without sensuous intuitions, or to use the famous Kant's dictum: "Thoughts without contents are empty, intuitions without concepts are blind" (*ibid.*, 45; A:51, B:75). Space and time, however, being forms of every intuition whatsoever, be it pure or empirical – when sensation (Empfindung) is present, are at the very "base" of sensibility (Sinnlichkeit). This is why space and time are referent to the knowing subject: they are *a priori* present forms, which make possible *the arrangement of all sensations in a phenomenon.* Only in this way do objects of experience become possible.

In a nutshell, these are the basic elements of Kant's aesthetical apriorism, concerning the rôle of space and time as necessary forms (accepting the matter) of every phenomenon. [8] Now, it is clear, I think, that Reichenbach and Carnap criticize not this original Kantian position, but some not well defined, yet naïve type of apriorism. Kant's apriorism is purely philosophical, while the criticized one is "naturalized" by introducing some innate tendency, compelling us to visualize the objects of the world as being placed in a three-dimensional Euclidean space. However, it is one thing to say that space and time are the *a priori* conditions for the very cognitive possession of objects, and another thing to insist that an innate property of the human mind is directing the way *how* to visualize objects, which might still conform to some different spatial geometry.

Notwithstanding his dissent from Kant's terminology, Hegel managed in a brilliantly didactical manner to demonstrate how space and time belong to consciousness through a figurative comparison, which I am tempted to adduce (and to which I will come back in the next section):

> There are things in themselves outside, but without time and space; and here comes the consciousness, which has in advance the time and the space in itself as a possibility of experience, just like to be able to eat, we have a mouth, teeth, and so on, as conditions for eating. Things, which are eaten up, do not have a mouth and teeth; and like eating concerns things, so the space and the time concern them; in the way eating puts the things between the mouth and the teeth, so consciousness puts them in the space and the time. (Hegel, 563, my translation[9])

So, space and time are owned by consciousness *as a possibility of human experience.* No *other* space and time are considered by the transcendental aesthetic. If so, then Kant could in no way confuse mathematical with physical geometry. The distinction (1), no matter how plausible, is not a Kantian problem. Consequently, thesis (2) proves to be an irrelevant critical argument against transcendental philosophy. *Kant has not made the error, attributed to him by Carnap and Reichenbach.*

Curious enough it might be, but not more relevant proves to be also the seemingly natural *explanation* of how the attributed error was possible, suggested by the otherwise benevolent critics. The explanation starts, as we know from the beginning of the section, with the fact that non-Euclidean geometry had not yet been discovered in the days when Kant was creating his philosophy. This is a good reason to maintain that "Kant should not be blamed for his error". Carnap convincingly also adds, that *it was not possible for Kant to think about geometry in any other way,* when "even mathematicians took this Kantian point of view for granted".

As a counter-example to his explanation I will point out, that Kant's theoretical insight was so vigorous, that it allowed him to suppose the possibility of spaces having other dimensionality, and with properties different from those of the perceptual flat three-dimensional space, *a long time before the discovery of non-Euclidean geometries.*

Already in his first work *Gedanken von der wahren Schätzung der lebendigen Kräfte* (1746), referring to his pre-critical period, Kant sees a basic connection between the existence of precisely three spatial dimensions and the law of change of the universal force of attraction – Newton's inverse square law of gravitation. According to him, this well-known law could be of some other kind. Instead of it – Kant speculates – God could have chosen another law, such as, for instance, that the force of attraction is inversely proportional to the cube of the distances among material bodies.

> An extension with other properties and dimensions would also spring from the other law. The science about all these possible kinds of space would be the highest geometry,

> which a limited understanding could undertake... If it is possible for extensions with other dimensions to exist, then quite probably God has indeed situated them somewhere. (Kant, *Gedanken*, 25, my translation)

I can safely say, that the quoted lines can be qualified as a bold prophecy for the birth of "the science about all these possible kinds of space", of the science about "*extensions with other properties and dimensions*", or in other words, for the birth of contemporary mathematical and cosmological theories about various spatial structures (with different metrics and topologies), and their application for representing regions of the universe. Kant's hypothetical reasoning was demonstrated, at that, before the discovery of non-Euclidean geometry in the nineteenth century, and before Ehrenfest's conclusions in 1917 that the stability of planetary orbits and atomic structures is essentially dependent on the dimensionality of space.

It is thus shown that we must attribute to Kant the merited respect for his hypothesis, rather than accept Carnap's remark, that it was not possible for him to think about geometry in any other way. Or, to make use in this sense of the frank astonishment of J. B. Stallo:

> It is remarkable how many of the scientific discoveries, speculations and fancies of the present day are anticipated or at least foreshadowed in the writings of Kant. (Stallo, 307, fn. 2)

4. KANT AND THE REALIST VIEW

The brilliant hypothesis about the possibility "extensions with other dimensions" to be realized somewhere, was suggested by Kant in his pre-critical creative period. In his later transcendental aesthetic, however, and most consistently in his first *Critique* and the *Prolegomena*, Kant ceases to refer to "space" and "time" as kinds of physical extensions, but denotes by these concepts the pure forms of sensibility. Here lies, to my mind, a subtlety in the interpretation of Kant's philosophy, to which I would like to draw special attention. It could serve, if not for a direct criticism – which, as I tried to show, is irrelevant – as an *indirect criticism* of the fact that Kant has turned his back to the possibility for space and time to be considered, if I use contemporary epistemological jargon, as theoretical constructs.

The mentioned subtlety stands close to a theoretical argument, which though not so popular, has already taken its place in the debates among connoisseurs and critics of Kant's philosophical heritage. I mean the so-called *neglected alternative argument* (NAA) against the well-known thesis of Kant, that space and time do not represent things in themselves in their relation to one another. It is ingeniously formulated by Hegel, in the adduced quotation from the previous section, commencing with the declaration that things in themselves outside are "without time and space". This thesis of Kant about the non-spatiality and non-temporality of things in themselves has entered specialized literature under the name "the non-spatiotemporality thesis" (Falkenstein). According to NAA, the non-spatiotemporality thesis is at variance with the assumption that Kant's conception of space and time as forms of human sensibility does not cancel the (neglected) alternative *things in themselves, out of any experience, to exist in some "transcendentally real" space and time of their own.* The latter are not perceived by us, and have nothing to do with the way in which the spatial and temporal arrangement of *phenomenal* items in sensuous intuitions comes about. The assumed transcendent space and time, as a matter of speculative suggestion may, or may not happen to resemble experienced space and time.

> The prime proponent of the neglected alternative argument was the 19th Century Kant-scholar, Adolf Trendelenburg, though Vaihinger has traced the objection as far back as Kant's contemporary, Hermann Andreas Pistorius. The neglected alternative argument was one of the points of contention in the protracted and acrimonious controversy between Trendelenburg and Kuno Fischer. (Falkenstein, 267-8, fn. 6)

In their nice book aiming to explain and defend Kant's critical philosophy, prepared especially for English readers (in a new edition in 1889), J. Mahaffy and J. Bernard succeed in launching a purely Kantian-like criticism on Trendelenburg's NAA. They congently point out that because there is no possible intuition that we can have apart from space and time, the latter are *empirically real*, but because of their *transcendental ideality*, if we omit the conditions of experience, they turn into nothing, and we have thus no reason to assert that they pertain to things in themselves.

> When Trendelenburg says that Kant forgot to inquire whether space and time might not be *both* subjective and objective, as a third possibility, he shows by his very statement of the question his ignorance of Kant's system. (Mahaffy and Bernard, 55, fn. 1)

I am not going to dwell on the historical discussions concerning NAA. The reason to leave aside its ontological pretentions is that this argument has solely an abstract speculative meaning. NAA does not properly concern the grounds for the *realist idea* about the applicability of definite kind of space to the physical world, within the conditions of given types of experiments, or more generally, practices. On the contrary, NAA considers the vague idea that some transcendent determinations *could be* isomorphic to the perceptual space and time in human experience. The speculative character of the argument, in my opinion, comes from the impossibility of such "transcendentally real" space and time to be transformed into an object of experience by a "standard" knowing consciousness. L. Falkenstein has made a successful attempt to show that the sense in which the objects of experience are ordered in space and time "is entirely distinct from any sense in which things may be supposed to be in space and time in themselves, independently of how they are presented in our experience" (Falkenstein, 283). We can neither affirm, nor deny, that things in themselves abide in a "space and time"of their own.

The cognitive "deficiency" of NAA, the fact that it is void of empirical content, does not make, however, contemporary scientists and philosophers more favourable to Kant's conception of space and time in his transcendental aesthetic. Their interest is not provoked by space and time in their quality of some transcendent determinations of noumenal essencies, but as it seems, they shun their definition as *a priori* intuitive forms, as well. The prevailing interest of physicists today in space and time, or more correctly, in space-time, is an interest in something that could be investigated by the methods and means of theoretical and experimental science.

Thus for example, already in the 1940s, S. I. Vavilov declared that "the new physics has transformed space and time from metaphysical categories into fundamental material entities" (Vavilov, 15); and among the claims of H. P. Robertson, stated almost at the same time, [10] that

> for Kant... the problem of physical space is solved on the epistemological, pre-physical, level.
> But the discovery of other congruence geometries, characterized by a numerical parameter *K*, perforce modifies this view, and restores at least in some measure the objective aspect of physical space, [11]

would probably typify most of the contemporary methodologists.

How could expressions like "the objective aspect of *physical* space" be interpreted from the point of view of Kant's transcendental philosophy? The answer to this question bears the already mentioned interpretative subtlety that I am now going to consider.

As I have already noted in the previous section, Kant (in his critical period) does not face the problem about the description of space and time as being outside the consciousness, i.e., about their description as some objective entities, or objective relations, standing against the cognizing subject. Does Kant's transcendental conception appear then to be merely *incommensurable* with the realist approach of today's scientists, and most philosophers of science, towards space-time?

The question is rhetorical, and its answer, as it seems, is clear. The metaphysical position of Kant does not share any conceptual touching point with the view of space-time as an objective, independent of the cognitive act, determination. In this case, if the new physics overturns the standpoint for the understanding of space and time, and, as S. I. Vavilov has put it, transforms them from metaphysical categories into fundamental material entities, then we are only a step away from the conclusion of the incommensurability of the two conceptions – Kant's and the realist one. The reason not to make the last step in accepting this conclusion is the following. I think that in the framework of Kant's transcendental approach, the possibility that space and time may be considered as specific phenomena of experience as well, the knowledge of which is verifiable, is not excluded on principal grounds. I would like immediately to point out, that *this possibility does not suppose the referents of the newly suggested notions to be transcendent determinations* – something which Kant would reject in a categorical manner, but which, as we have already seen, is the central claim of NAA. The suggested space and time would be neither qualities of things in themselves, nor would they represent them in their relation to one another, as Kant requires. However, there is no obstacle space and time to be additionally introduced as auxiliary notions, expressing the specific setting of objects of experience. In the contemporary (essentially post-Kantian) methodological vocabulary such notions are called "pure *theoretical* concepts", or shortly, "theoretical constructs".

My suggestion that space and time could also be "objectified" in a realist mode, by the possibility to be looked upon as theoretical concepts, is at a first glance, matching with the speculative pretention of NAA. At a second glance it is seen that the suggested possibility has another, and far less ambitious, cognitive pretention. In support of this possibility, though it might appear to be a forced "modernising" of transcendental philosophy, different examples of theoretical concepts can be adduced, which Kant himself has readily admitted in their quality of legal "participants" in theoretical knowledge.

One such concept is that of the magnetic field (or, at the time of Kant, of the invisible "magnetic matter"). When considering the postulates of empirical thought, Kant speaks about the existence of magnetism, which cannot be directly registered by our sensory organs, but which is introduced to explain the observed phenomena:

> Thus we know the existence of some magnetic matter pervading all bodies from the perception of the attracted iron filings, though our organs are so constituted as to render an immediate perception of that matter impossible. (Kant, *Critique*, 175; A:226, B:273)

When elucidating the principle of the anticipations of perception, Kant involves illustrations concerning complex theoretical concepts, like the *degrees* of heat expansion (*ibid.*, 143, A:174, B:216). In the preface to the second edition of the *Critique of Pure Reason* (1787) he writes about the establishment of the hypothesis, assumed by Copernicus, about the invisible force of attraction – Newton's force of gravitation – "which holds the universe together" (*ibid.*, XXXV; B:XXIII).

All three phenomena – the magnetic field, the heat radiation, and the force of gravitation, have been presented in the history of science by theoretical constructs, based on mathematical apparatus with enhancing complexity: separate field, Maxwell's differential equations, quantified field, direct force at a distance, curvature of Reimannian space-time, components of the metric tensor, and the like.

The mentioned theoretical concepts, constructed by reason, but explaining given types of experimental observations, are quite admissible by transcendental philosophy, because the phenomena described by them co-ordinate with the rest, in compliance with the analogies of experience. However, if so, why can a theoretical concept of space not be fully admissible, as well. If the new concept of space would jar in the ear of orthodox kantians, it can be altered, for example, into the expression "physical medium of dynamic phenomena". Nevertheless, the purpose of its introduction is to facilitate the theoretical description of the motion of bodies and of dynamical processes in different sorts of experiments. Namely with this "innovation" transcendental philosophy could assimilate those hypothetical cases of different physical extensions we have already seen the "early" Kant to suppose in his *Gedanken*. The theoretical concept of space may also be of great help for the substantiation of Kant's dynamical definition of the concept of matter in his much later work *Metaphysische Anfangsgründe der Naturwissenschaft* (1786), by giving theoretical meaning to the notions of filled up space (erfüllter Raum), and of empty space (leerer Raum).[12]

The possibility for the introduction of concept(s) of space as theoretical construct(s), by leaving at the same time transcendental aesthetic intact, should in no way be interpreted as a direct criticism of Kant, but at the most as a kind of indirect critical argument. At that, only to an extent, to which Kant was able to predict, that the growth of scientific knowledge could propel the thinking reason to an assumption, which, for Kant himself, has stayed far aside transcendental philosophy. An assumption that looks quite natural today for philosophers of science, viz. *the idea about the empirical testability of the concepts of space and time (space-time), as concepts of physical geometry.*

Kant was not bound to deduce this, even if he had anticipated this possibility. At least for me, what is more important is the claim itself, that the transcendental approach does not raise principle barriers against the possibility for theoretical constructs of space (space-time). When this possibility fails to be noticed, even prudent philosophers address hard critical remarks towards Kant. Thus for instance, Milič Čapek tries to convince us of the fact that Kant

> would be equally shocked by the Reimannian geometry as by Lemaître's and Gamow's cosmogonies assuming the finiteness of the cosmic past bounded by the initial 'zero-time'. (Čapek, 43)

I dare say, on the contrary, that Kant's bold ontological and epistemological hypotheses (about "extensions with other properties and dimensions", and "the science about all these possible kinds of space"), considered in the end of the previous section, are rather an indication that Kant might not have been so shocked by the mentioned physical and cosmological novel theories. There is an egress for the kantian philosopher: the possibility remains open for operation with theoretical constructs of space and time, while keeping the basic position of transcendental aesthetics, in which space and time are not defined as (empirical or theoretical) concepts, but as pure *a priori* intuitions.

A serious perplexity, however, comes to the fore: the awkwardness, or inconvenience because of the conceptual (and semantical) cleavage of the terms "space" and "time". Quite probably, orthodox kantians would turn their back to the suggested possibility, which would only overburden the orderly conception of Kant. The indirect criticism, addressed to him, that he had not foreseen the future development of physics and cosmology, can prove to be more acceptable than a conceptual complication by idealisations, which (though not contradictory) are not akin to the basic transcendental meaning of space and time. What then?

No other alternative is then left for us, except *to make the last step towards the consent with the incommensurability* between Kant's transcendental philosophy and the general realist view of space and time. An alternative, no matter how unacceptable it may appear, is twice as protecting of transcendental philosophy from direct criticism.

NOTES

[1] Cf. (Florensky, 7).
[2] See M. Čapek's Introduction in (*The Concepts*, XXXVI). My italics.
[3] (Grünbaum, 154). I'll analize Reichenbach's "telling contribution" in the next section.
[4] (Carnap, *Foundations*, 181). My italics.
[5] The distinction between *mathematical* and *physical geometry* is cogently explained by Carnap in his book – see *ibid.*, pp. 181-183.

[6] Cf. *Ibid.*, ch. 18. He also clearly states elsewhere that "It is the contention of construction theory that there are no such things as the "synthetic judgments a priori", which are essential for Kant's approach to epistemological problems." (Carnap, *Logical Structure*, 176)

[7] Theorem: "Given a geometry *G*' to which the measuring instruments conform, we can imagine a universal force *F*, which affects the instruments in such a way that the actual geometry is an arbitrary geometry *G*, while the observed deviation from *G* is due to a universal deformation of the measuring instruments." (Reichenbach, 33)

[8] The all-pervasive rôle of these forms can be seen when the answer is being sought to the simple question: *why are non-Euclidean figures not possible to be perceived*? The answer is successfully discussed by David Sherry who turns the philosophical attention to Kant's image forming synthesis, which is one and the same both for geometrical constructions and for perceptions of empirical objects. He clearly brings out Kant's aprioristic position, that "we construct objects of experience according to the same principles by which we construct mathematical objects" (Sherry, 35). It becomes clear now why non-Euclidean figures are ruled out as impossible perceptions: "the same formative synthesis, by which we construct a triangle in imagination, should be identical with that which we exercise in the apprehension of a phenomenon, in order to make an empirical concept of it." (Kant, *Critique*, 174; A:223, B:271)

[9] I thank Valentin Kanawrow for a more precise comment on the translation.

[10] H. P. Robertson's paper was published for the first time in the well-known book *Albert Einstein: Philosopher Scientist*. P. A. Schilpp (ed.), New York: Harper &Row, 1949.

[11] (Robertson, 414). "The numerical parameter *K*" signifies the curvature, as a numerical function of place, of the non-Euclidean space.

[12] "*Matter* is the mobile, as far as it *fills up* a *space*. A space to be *filled up* means anything movable to be resisted, which, through its motion tries to penetrate a given space. A space, which is not filled up, is an *empty* space." (Kant, *Metaphysische Anfangsgründe*, 227, my translation)

REFERENCES

Čapek, Milič. *Bergson and Modern Physics. A Reinterpretation and Re-Evaluation*. Dordrecht: D. Reidel, 1971.

Carnap, Rudolf. *Philosophical Foundations of Physics. An Introduction to the Philosophy of Science*. New York, London: Basic Books, Inc. Publishers, 1966.

Carnap, Rudolf. *The Logical Structure of the World and Pseudoproblems in Philosophy*. Berkeley and Los Angeles: University of California Press, 1967.

The Concepts of Space and Time. Their Structure and Their Development. Milič Čapek (ed.), Dordrecht: D. Reidel, 1976.

Curd, Martin and J. A. Cover. *Philosophy of Science: The Central Issues*. New York, London: W. W. Norton & Company, 1998.

Falkenstein, Lorne. "Kant's Argument for the Non-Spatiotemporality of Things in Themselves." *Kant-Studien* 80. Jahrg., Heft 3, 1989: 265-283.

Florensky, P. A. *Analysis of Spatiality and Time in Works of Pictorial and Plastic Arts* (in Russian). Moscow: Publishing Group "Progress", 1993.

Grünbaum, Adolf. *Philosophical Problems of Space and Time*. Dordrecht: D. Reidel, 1974.

Hegel, G. W. F. *Vorlesungen über die Geschichte der Philosophie*, in: Hegel, G. W. F. *Sämtliche Werke*. Hg. Von Hermann Glockner, B. 19. Stuttgart: Fr. Frommanns Verlag, 1941.

Hramova, V. L. *Philosophical Analysis of the Problem about the Relation between Theory and Experiment in Relativistic Physics* (in Russian). Kiev: "Naukova Dumka", 1974.

Kant, I. *Gedanken von der wahren Schätzung der lebendigen Kräfte*, in: Kant, I. *Sämtliche Werke*. Hg. Von R. Vorländer, B. VII (49). Leipzig: Felix Meiner Verlag, 1922.

Kant, I. *Critique of Pure Reason* (translated by F. Max Müller). Garden City, New York: Anchor Books, Doubleday & Company, Inc., 1966.

Kant, I. *Metaphysische Anfangsgründe der Naturwissenschaft*, in: Kant, I. *Sämtliche Werke*. Hg. Von R. Vorländer, B.VII (48a). Leipzig: Felix Meiner Verlag, 1922.

Karapetyan, A. A. *Critical Analysis of Kant's Philosophy* (in Russian). Erevan: Armenian State Publishing House, 1958.

Mahaffy, John P. and John H. Bernard. *Kant's Critical Philosophy for English Readers*. London and New York: Macmillan and Co., 1889.

Popper, Karl R. *Conjectures and Refutations. The Growth of Scientific Knowledge*. London: Routledge, 1989.

Robertson, H. P. "Geometry as a Branch of Physics", in: *The Concepts of Space and Time*: 409-424.

Sherry, David. "Construction and *Reductio* Proof." *Kant-Studien* 90. Jahrg. Heft 1. 1999: 23-39.

Stallo, J. B. "Criticism of Newton, Euler, Kant and Neumann", in: *The Concepts of Space and Time*: 305-307.

Vavilov, S. I. "Main Roads of Contemporary Physics", in: Polikarov, A. *Achievements of Contemporary Physics* (in Bulgarian). Sofia: Narizdat, 1945.

Zucker, Francis J. "Supplementary Comments to Weizsäcker's Paper", in: *Bost. Stud. Phil. Sci.*, Vol. V Dordrecht: D. Reidel, 1969.

DIMITRI GINEV

HOW TO BE SIMULTANEOUSLY AN ANTIESSENTIALIST AND A DEFENDER OF SCIENCE'S COGNITIVE SPECIFICITY

1. INTRODUCTION

Reflecting upon the issue of science's cognitive specificity in the framework of the analytical philosophy of science helps one to draw a clear demarcational line between the two traditional contexts of scrutinizing scientific research. Studies in the context of justification advocate this specificity by postulating a cognitive essence (e.g., scientific method, invariant standards for rational scientific behavior, primitive language of the "protocol-sentences", reticulated network of methodological norms and cognitive values, and the like). The postulated cognitive essence makes science a "natural kind" and thereby, "the science-nonscience distinction somehow cuts culture at a philosophically significant joint." (Rorty, 50) By contrast, historians, sociologists, psychologists, ethnomethodologists, and other practitioners of science studies in the context of discovery claim that all alleged cognitive essences distinguishing science as an intellectual enterprise are only socially produced contingent and contextual formations. As a consequence, the opinion began to gain currency that because cognitive essentialism is wrong, there is no cognitive specificity of science. The more one is engaged in investigating science by following the way from its "final products to production, from 'cold' stable objects to 'warmer' and unstable ones" (Latour, 21), the more one is inclined to deny that there is "something special" about scientific Method and Rationality. According to those who reject any form of cognitive essentialism, only by treating science as a black box (i.e., by discarding the discursive-practical texture of routine scientific research) there arises the illusion of an irreducible cognitive specificity. And vice versa, the illusory epistemological uniqueness of "ready-made science" vanishes when one looks at "science in the making", where (to make use again of a dictum raised by Latour) "context and content merge".

Thus, within both traditional contexts (of discovery and justification) the coupling of antiessentialism and the defense of an irreducible *differentia specifica* of science proves to be an "impossible combination". To defend science's cognitive specificity without succumbing to cognitive essentialism requires designing a new

Dimitri Ginev (ed.), Bulgarian Studies in the Philosophy of Science, 187-206.

context of science studies. A concern to attain such a context harks back to a time when authors like Joseph Kockelmans, Theodore Kisiel and Patrick Heelan raised the claim that the hermeneutic aspects of scientific research belong neither to the psychology and sociology of research work nor to the "rational reconstruction" of ready-made knowledge. What is important in this regard is that the investigations of these authors traffic primarily in a methodology that I should like to call a "cognitive existentialism". This is a radical alternative to cognitive essentialism.

The aim of the present essay is to prove whether cognitive existentialism (qua a methodology that resists classification within the traditional discovery-justification dichotomy) does provide arguments for defending science's cognitive specificity. In Section 2, I discuss objectifying thematization as a principal distinctive feature of science. I shall make some observations about the attempts to trivialize objectifying thematization by reducing it to a decontextualization of the outcome of scientific research. Section 3 is devoted to the concept of cognitive essentialism. After some preliminary methodological remarks, I focus on the four philosophical programs promoting the contemporary variants of cognitive essentialism. Section 4 tries to articulate the main ideal types of cognitive essentialism. I first differentiate between five principal aspects of the essentialist approach to science's cognitive specificity. This differentiation enables one to determine the relations between the aspects and the variants of essentialism, which provides the opportunity to single out the ideal types. My aim in Section 5 is to shed light on the antiessentialist strategies that fail to oppose cognitive essentialism in a radical manner. I explicate *radical antiessentialism* as a conception that opposes all principal aspects of the essentialist understanding of science. In Section 6, the possibility of *cognitive existentialism* as a radical alternative to cognitive essentialism is analyzed. By focusing on the hermeneutic and constitutive (in the sense of phenomenological philosophy) dimensions of scientific research, I shall finally (in Section 7) try to suggest a hermeneutic-existentialist defense of science's cognitive specificity by scrutinizing objectifying thematization in the context of the new strategy.

2. SCIENCE'S COGNITIVE SPECIFICITY

My contention is that the central indisputable distinctive feature of science is *objectifying thematization.* In each and every scientific domain the research process leads to a thematization of objects of inquiry.[1] The very thematization is not a static relation between a "thematizing subject" and a "thematized reality". It is rather a constructive process in which one "makes present" objects of inquiry. By "objectifying thematization" I mean in the first place *the way of "making present" by constructing theoretical models in the research process.* In another definition, objectifying thematization is a project of disclosing entities in an objective way, i.e., as systems of data predicated on the possibility to be theoretically interpreted (or interpreted by means of theoretical models). Speaking in terms of Heidegger's "existential conception of science", the emergence of objectifying thematization lies in the interrelatedness of discursive practices, which promotes at once a releasing of domains of research from the pre-theoretical "world of routine everyday practices" and a mathematical projection of the structures of these domains. The releasing of themes from the everyday world and the mathematical projection imply that objectifying thematization operates in "mathematically homogeneous worlds" composed by the models of the objects of inqury. (Thus, the "possible worlds" of objectifying thematization can be investigated in terms of a philosophical conception of the ways of constructing and interpreting theoretical models.)

Speaking in more technical terms, objectifying thematization involves four moments: (a) formulating principles of invariance and groups of symmetry operations; (b) elaborating on covariant differential equations; (c) constructing theoretical models united by a common state-space where each model assigns to a certain system of data a trajectory in that space[2]; (c) devising explanatory scenarios through interpretation of the models. Objectifying thematization delineates a domain of research through stating principles of invariance (for instance, invariance with respect to displacement in time and space, or invariance that can be derived from the equations of electrodynamics, or the principle that all actions are transmitted by fields which transmit the perturbations from point to point).[3] By means of these principles one determines the necessary conditions that the equations of theory's models in a domain of research must satisfy. These equations inform the state-space characterizing the domain's systems of data. In specifying this state-space, each particular theoretical model presents a thematization of the dynamical behavior of a certain system of data. From a semantic point of view, a theoretical model objectifies the dynamical behavior as a set of states and a sequence defined over that set. Each state is a simultaneous configuration of values of the basic behavior's parameters. For instance, a state of a system studied in classical mechanics is a configuration of the instantaneous positions, masses, and velocities of the bodies included in the system. (Only a small number of scientific domains are constituted by theoretical models that are not dealing with systems' dynamical behavior but with taxonomical, morphological, or structural aspects of empirical systems. As a rule, however, these domains are closely related to domains in which systems' dynamical behavior is investigated.)

The sociologists of scientific knowledge are inclined to replace objectifying thematization by what they call a "decontextualization" of the outcome of the research process. On their view, decontextualization is to be investigated only on the level of preparing scientific papers. It refers to the techniques of purging of personal motifs and situational contingencies in the text's design. For sociological study, this kind of textual decontextualization means at the same time a recontextualization, by inserting the outcome of a scientific work into a "web of reason", i.e. the lines, of argumentation surrounding a research effort that provide reasons for the selections of resources, interests, methods and so on.[4] On various programs of social constructivism, the translation of objectifying thematization into the vocabularies of de- and recontextualizaion in designing the texts of scientific papers is a completely sufficient procedure for explaining the production of scientific knowledge.

Closely related to the replacement of objectifying thematization with procedures of decontextualization is another argument against the possibility to "distillate" a cognitive specificity of science. In this argument, the more a scientific domain attains an esoteric content (and a higher degree of cognitive autonomy distinguished by the specialized formal methods and techniques of thematization) the further it extends outside in the sense of adding more and more outside people to the making of science. Those who maintain a domain's cognitive autonomy are only a tiny group among the huge army of people who do science. In view of the expanding networks including more and more non-cognitive settings and factors of the "social manufacture" of scientific knowledge, the search for stylizing features of science's cognitive specificity does not seem a reasonable undertaking.

In a tentative manner, I should like to stress that the main fault of the sociologists' attempts at deconstructing science's cognitive specificity is the overlooking of the textures of discursive practices that generate domains of objectifying thematization. What the sociologists of scientific knowledge fail to take into account is the projection and constitution of domains of research. The stability of these domains, despite the fact that the production of scientific knowledge is embedded in quite heterogeneous settings, cannot be traced back to the de- and recontextualization of scientific papers. *"Projection" and "constitution" are processes that cannot be grasped in terms of a social construction of scientific knowledge.* Furthermore, the social constructivists ignore the peculiar horizons of doing research informed by the interrelatedness of such practices. They are only interested in relativizing the particular practices to different social contexts, ignoring thereby the "cognitive reality" arising within the discursive-practical horizons of doing research.

Let me return to the main issue of this section, the role of objectifying thematization as a paramount trait of science's cognitive specificity. To carry out such a thematization presupposes an extended cognitive structure of scientific research. There is no objectifying thematization that is not related to a normative methodology. The explication of objectifying thematization cannot proceed without criteria and standards of truth. It is impossible to make the thematization of objects of inquiry by means of theoretical models explicit without epistemic criteria of objectivity. Obviously, the efforts of elucidating objectifying thematization focus on the diverse cognitive essences one is using in the context of justification. One also makes use of these essences when one is trying to draw a philosophically important demarcational line between science and the rest of culture. Yet the fact that they are indispensable for making objectifying thematization explicit does not entail that normative methodology, standards of truth, criteria of objectivity and the like are a priori conditions for carrying out objectifying thematization. Rather, these essences are "created" within the processuality of objectifying thematization. *Vis-a-vis* the fact that they are not "independent determinants" of scientific research, we must now ask the following question: Can one give an account of objectifying thematization by avoiding any appeal to cognitive essences? A positive answer to this question is the first step in defending science's cognitive specificity from an antiessentialist point of view. Before handling the possibility of approaching objectifying thematization in "existentialist" terms, however, I should like to bring into focus the complexity of the concept of cognitive essentialism.

3. ON THE CONCEPT OF COGNITIVE ESSENTIALISM

By cognitive essentialism I mean, roughly speaking, the view that in scientific knowledge (and only in this knowledge) an invariant (ahistorical) essence of human cognition takes place. There are two corollaries to this view. First, there is no pre- or non-scientific form of knowledge that can incorporate this essence. (Phrased in a quasi-Hegelian manner, the absolute essence of human cognition has had to await all genealogical steps of human intellectual history in order to "attain itself" in science as the final genealogical step.) Second, the invariant cognitive essence is a "Platonic entity", since it is supposedly independent of (i.e., not step-by-step derivable from) particular discursive practices and their "temporal configurations".

As I pointed out, according to the dominant account, to stress the view of cognitive essentialism is a prerequisite for defending science's cognitive specificity. Although this account within the non-analytical (Continental) approaches to scientific research has increasingly been put in question, the majority of the philosophers of science still believe that only by reconstructing an epistemological and/or methodological and/or metaphysical and/or transcendental "essence" of science, can one defend its cognitive specificity in a coherent manner. By implication, depending on the philosophical framework of supporting cognitive essentialism, the specificity of science can be cast in terms of (i) normative epistemology or (ii) naturalized epistemology or (iii) scientific realism or (iv) a kind of transcendental philosophy. Since scientific realism is (in principle) reducible to a kind of naturalism and the transcendental strategy to cognitive essentialism leads to a normative epistemology, the essentialist understanding of science can basically be spelled out in the frameworks of (i) and (ii). The case (i) is most typically illustrated by Lakatos' rational reconstructions of the history of science based upon different "codes of scientific honesty". In case (ii), by using Quine's "replacement thesis", science's cognitive essence is defined in the frameworks of scientific disciplines like AI, cognitive psychology and evolutionary biology. This leads inevitably to a kind of paradoxical self-authentication of science's cognitive specificity, since the cognitive essence that determines the *differentia specifica* of science is defined in terms of a particular scientific theory. To be sure, similar difficulties arise also with justifying the normatively postulated cognitive essence of science. More specifically, by admitting a normative version of cognitive essentialism one is unable to meet such "classical" objections like the descriptive/normative paradox and the so-called Fries' trilemma.[5] However, despite these difficulties, to defend cognitive essentialism in terms of normative epistemology seems (at least intuitively) a much more consistent strategy than defining the "cognitive essence" of science in terms of science itself. This is why in the rest of this essay my attention will be concentrated first and foremost on the normative versions of cognitive essentialism.

4. IDEAL TYPES OF COGNITIVE ESSENTIALISM

I would like now to distinguish between five principal aspects of defining and defending science's cognitive specificity in terms of essentialism. Each strategy of essentialist interpretation of science dwells on one or more of these aspects.

(a) The first aspect concerns cognitive essentialism in terms of *scientific method.* Although the view that scientific research is governed by a uniform and invariant scientific method is no longer to be defended, the belief that behind the plurality of methodologies one can find invariant (meta)methodological principles and norms is still alive. The argument for the indispensability of such principles and norms runs as follows. If there is no invariant set of (meta)methodological principles (or, an ultimate "objective viewpoint") the idea of scientific progress collapses into a kind of cognitive relativism.[6] In other words, despite the growing proliferation of scientific methodologies, one needs a (meta)methodological invariance if one wants to provide a rationale for the persisting "system of reference" from which the progress of scientific knowledge is to be evaluated. The invariant set of (meta)methodological principles that persists in the various situations of doing research is the most celebrated aspect of the essentialist image of science.

(b) The second aspect is cognitive essentialism in terms of the *objective truth of science.* In a traditional view, truth as a correspondence to reality is attributed to particular propositions of scientific knowledge. These propositions acquire a special status within the whole body of knowledge, because they reveal the "ultimate structure of reality". With regard to this posture, there are two possible ways of defining objective truth as a cognitive essence of scientific enterprise. The first way was already indicated. It consists of the localization of objective truth to a set of particular propositions within the structure of scientific knowledge. The second way is oriented towards the use of concepts that transcend the intrinsic conceptual frameworks of science. The localization of objective truth is employed (for example) by the logical positivists, while the latter is preferred by the scientific realists.

(c) The third aspect is cognitive essentialism in terms of the *objectivity of scientific knowledge.* By stressing this aspect, one claims that scientific knowledge has a privileged access to reality. There is a "core of privileged representations" that builds up the "bottom" of scientific knowledge.[7] To be sure, stressing this aspect does not imply (by necessity) a commitment to a realist position and/or to a correspondence theory of truth. One can hold that all epistemic relations to reality are historically, socially and culturally constructed, nevertheless, among these relations science's epistemic relation to reality is (according to certain epistemological criteria) most reliable. Or, one can admit that scientific truth depends in a crucial manner on the mechanism of achieving consensus in a scientific community (subscribing thereby to something like a consensus-cum-pragmatic

picture of truth), nevertheless, one can still hold that scientific knowledge provides a privileged access to reality. What, however, one is by necessity assuming when stressing the privileged epistemic access to reality is the representational relation of a certain class of propositions in scientific knowledge to the "definite world structure". It is this relation that takes the form of a characteristic cognitive essence of science. Thus, defending cognitive essentialism in terms of objectivity of scientific knowledge has much to do with epistemological foundationalism. I would say that the former is a strategy of arguing for the latter within philosophy of science. (A further elaboration of this aspect leads to the view that the scientific description of reality constitutes a "privileged taxonomy", since only it can reveal how reality breaks into "natural kinds". In so doing, one goes on to extend epistemological foundationalism to a kind of "objectivist metaphysics of science".) To define the objectivity of scientific knowledge as a cognitive essence of science requires not only an appropriate epistemology but also the framework of a theory of meaning. This can be a traditional theory, which stipulates that by choosing certain properties as defining a type of thing (or a "natural kind"), we give the meaning of the term we are using for that type. Or, it can be a "causal theory of reference" that in contrast to the traditional theories advocates the historical permanence of meaning (of a given type of thing) despite any changes in the descriptive criteria associated with the thing. As Shapere points out, both kinds of theory of meaning (and reference) stem from the idea that one can lay down inviolable conditions on the knowledge-seeking enterprise by an analysis of semantic features of language that are supposedly "linguistic essences". Meaning and reference are "ideal entities" whose semantic and epistemological specifications have to serve as tools in the attempt to define the objectivism of scientific knowledge as a cognitive essence.

(d) The fourth aspect is cognitive essentialism in terms of the *rationality of science*. To be sure, there are many sub-aspects under this heading. In the present discussion, I refer to "scientific rationality" as a cognitive essentialism that defines epistemological criteria of demarcation between science and nonscience. Of course, there is a variety of sociological and cultural-theoretical crietria for demarcating science from the other institutionalized human activities, which reproduce themselves culturally and historically by means of discursive practices guided by epistemic criteria and norms. Yet, maintaining that science has an intrinsic rationality that in a philosophically significant manner distinguishes the structure of scientific knowledge and the cognitive organization of scientific research from the cognitive procedures one uses in the non-scientific discursive practices, requires (according to this aspect of cognitive essentialism) formulating a demarcational criterion in epistemological terms. Popper's falsifiability criterion (as defined in the framework of the "epistemology without knowing subject") is an appropriate illustration in this regard. To sum up, the epistemological distinctiveness of scientific rationality is an aspect of science's cognitive essence. (Of course, most philosophers of science would claim that the aspect (*d*) does collapse into aspect (*a*). However, one need not necessarily seek to understand the rationality of science in terms of the method it employs. Thus, for example, scientific rationality can be

defined in terms of the optimal structure of scientific theory, or in terms of following the best cognitive aims and goals.)

(e) The final aspect is cognitive essentialism in terms of the *privileged status of science in modern societies*. Here I am referring to what Joseph Rouse calls "legitimizing narrative" of how the cognitive essence of science contributes to the technological (as well as economic and institutional) progress of modern societies.[8] When one thinks of a final legitimation of the privileged political and cultural status science enjoys (a status described by Feyerabend in terms of the "twentieth-century ideology and mythology"), one typically provides a "narrative" that relates a supposed cognitive essence of scientific research to the basic achievements of modern civilization. And vice versa, by defending the "project of modernity" and the rationality of modern civilization, one often takes recourse to the cognitive essence of science. In so doing, one "tells a large story" about the important cultural-historical events related to the cognitive organization of scientific research. The possibility to legitimate narratively the privileged political and cultural status of science in modern societies is a dimension of science's cognitive essence.

Depending upon which of these aspects more emphasis is placed, one is to distinguish between five "ideal types" of cognitive essentialism. Each of them covers a large set of "real variants". It is not my intention to enter here into a discussion of the possible variants of cognitive essentialism. What is of significance in the present context is a basic distinction between non-radical and radical versions of this view. Only by taking into account all five aspects could one maintain a radical version of cognitive essentialism. *Ex hypothesi*, by excluding one or more of these aspects the view of cognitive essentialism will be compatible with a given kind of antiessentialism. In other words, each partial (non-radical) version of cognitive essentialism will at a certain point no longer contain the resources to withstand the "invasion" of antiessentialist arguments. One place in which the compatibility (and in some cases even the complementarity) between variants of cognitive essentialism and variants of antiessentialism has emerged is the realism/antirealism debate in the philosophy of science.

5. NON-RADICAL VARIANTS OF ANTIESSENTIALISM

Depending on which of the constitutive aspects of cognitive essentialism is attacked, one could distinguish between five "ideal types" of antiessentialism. Let me briefly describe them.

By attacking (*a*) one arrives at a certain kind of cognitive relativism.[9] More specifically, this would be a view denying the possibility of having an "absolute point of reference" when evaluating scientific progress. Usually, this relativism concerning scientific progress is associated with Kuhn-Feyerabend's incommensurability thesis. Of prime importance here is the rejection of a "neutral experience" (i.e., on a meta-level with respect to the intrinsic methodological discussions in science) that can recognize a progressive shift in the development of science. Yet cognitive relativism is not in dissonance with the essentialist picture of science. Since it does not travel without the (essentialist) idea of conceptual scheme (and all epistemic and semantic dualisms promoted by it), cognitive relativism is rather a kind of non-objectivist, non-universalist and non-historically-invariant cognitive essentialism.

Refuting (*b*) is the main enterprise of all sorts of constructivism. In the contemporary philosophy of science, van Fraassen's "constructive empiricism" is the most celebrated program that illustrates this type of antiessentialism. In a succinct manner, van Fraassen (1989, 13) demonstrates that to study science's structure in and by itself, "as a product of the intellect that strives to order and unify the deliverances of experience", offers an alternative approach to the reification of "objective truth" that goes beyond the actual, observable phenomena obtained in scientific research. Van Fraassen's antiessentialism resists not only this kind of reification but also the demand for explanations of the regularities in the observable course of nature by means of scientific laws. Constructive empiricism as an alternative to the reification of scientific laws departs from the structure of theory's models. It only pays attention to structural features of theory, which delineate the structure of any possible world allowed by theory.

Usually, the attempts at discarding (*c*) are associated with attempts to figure out an alternative both to realist and antirealist positions in the philosophy of science. Fine's program of the "natural ontological attitude" is perhaps the most promising candidate for such a post-realist-antirealist position. All kinds of antiessentialism of this group are suggesting ways of how to "deprivilege" the special sort of science's epistemic access to reality that is built into the essentialist picture of scientific objectivism. It is the stress upon this epistemic access that forms the common denominator of realism and antirealism. Arthur Fine argues that what binds realism and antirealism together is the view of science as a set of practices in need of interpretration (i.e., an interpretation with respect to the aforementioned epistemic access), and that both realists and antirealists see themselves as providing just the right interpretation.[10] Discarding (*c*) provides the opportunity to demonstrate that science is not needed in this interpretation.

The aim of undoing (*d*) is to show that "scientific rationality is ... a pleonasm, not a specification of a particular, and paradigmatic, kind of rationality, one whose nature might be clarified by a discipline called 'philosophy of science'." (Rorty, 71-72) In admitting this antiessentialist strategy, one seeks to stress the lack of philosophically significant science/nonscience cut of the integral web of cultural meanings and values. The "ideological motif" that guides the variants of antiessentialism included in this group is the defence of the unity and integrity of the "whole of culture". Of course, Rorty's "epistemological behaviorism" is the most prominent example in this regard. Rorty's aim is to drop all traditional distinctions put forward by the epistemological theories of scientific rationality, and thereby to dispense with the attempts to mark off "the whole of science" from the "whole of culture". According to his neopragmatic antiessentialism concerning the issue of scientific rationality, all our beliefs and desires should be seen as part of the same Quinean web. (Rorty, 56) The success of science has to be explained not by appealing to a special rationality of scientific research. There is only a pragmatic criterion for such a success (and for separating science from nonscience) - some increase in our ability to predict and control that one can trace back to the development of science.

Finally, by attacking (*e*) one is trying to figure out either a "political philosophy of science" or a "politics of the postmodern philosophy of science".[11] The target of Rouse's antiessentialist line of argument is the view of cognitive essentialism as providing arguments for the "pro-modernist legitimation project". According to this kind of antiessentialism, since science is not predicated on having a "cognitive essence", there is only room for carrying out particular "cultural sudies" of the research activities. The core of Rouse's "politics of postmodern philosophy of science" is the "deflationary accounts of scientific knowledge". He starts out from the assumption that by relegating practices of scientific research to the background within which claims of knowledge acquire epistemological justification, one commits a mistake. In avoiding this mistake, which is associated in the first place with the postpositivist philosophical tradition, the deflationary accounts prove to be incompatible with other variants of antiessentialism like epistemological skepticism, relativism, and eliminativism (e.g., Steve Fuller's program of substituting the networks of social causation for the categories of epistemic rationality and justification). A deflationary account of scientific knowing treats it as lacking sufficient integrity and unity to support substantive generalizations about its instances. Thus, following such an account, one would be able to interpret and assess particular scientific claims but one would not have grounds for global interpretations and rationalizations or critiques of scientific research as such. (Rouse, *Engaging* 180) According to Rouse's basic belief, one bids farewell to an essentialist understanding of science, when one simply gives an account of scientific knowledge as located in the dynamics of practices of scientific research. The deflationary approach to scientific knowledge hinges on two views, which are compatible with versions of essentialism. On the one hand, Rouse assumes that the ways in which a construction of scientific knowledge is articulated and deployed in further research depend on considerations that do not respect internalist boundaries. Obviously, this assumption draws on the internalism-externalism distinction. This distinction is tacitly premised in Rouse's deflationary accounts. Yet it can only be accepted from an essentialist position. On the other hand, in rejecting any form of cognitive holism, the postmodern philosophy of science introduces a kind of "practical holism". Paradoxically enough, Rouse's extreme anticognitivism inaugurates a reificataion of (however dynamic they are) alignments of practices, environments, and agents situated within them. Thus, the "politics of the postmodern philosophy of science" replaces cognitive essentialism with a sort of "practical essentialism". As Vogel (1991, 271) rightly observes, by suggesting that there are practices "that will always escape theoretical scrutiny, Rouse is reintroducing the very notion of an a priori that anti-foundationalists such as Quine have taught us to reject."

Just as the plurality of variants of cognitive essentialism is not to be reduced to the five "ideal types" listed above, so the diversity of antiessentialist views is much greater than these five "ideal types", which I tried to illustrate with some prominent examples. Given the diverse array of relativist, contextualist, particularist, conventionalist, constructivist, pragmatist, and postmodernist views that have sought the antiessentialist label, it is impossible to identify the core of the contemporary attacks against cognitive essentialism. Yet a common denominator of the views covered by the five ideal types is that they oppose one or more particular aspects of cognitive essentialism. By implication, they do not suggest a radical antiessentialist alternative. Let me stress again that only by opposing the totality of the five central aspects of the essentialist understanding of science is one able to figure out a radical alternative.

With respect to the issue of science's cognitive specificity, there is an important similarity between (at least some of) the non-radical antiessentialist views and the different sorts of cognitive essentialism. Both parties (the naturalist and non-naturalist champions of science's cognitive specificity and those who reject that there is a significant science/nonscience cut in the whole of culture) agree that one needs cognitive essentialism in order to defend a *differentia specifica* of science. What binds believers (like Lakatos) in the internal logic of science's cognitive dynamics and postmodernists (like Rouse), is the presupposed one-to-one correspondence between cognitive essentialism and science's cognitive specificity. Yet, in both cases one is unable to meet the objection of vicious circularity. In the former case, the "proof of science's cognitive specificity" is always tacitly assumed because of the hypostatization of the internal logic of the history of science.

In the case of non-radical antiessentialism the assumption of one-to-one correspondence implies that the way of overcoming science's cognitive specificity is based upon a *petititio principii*. More specifically, "deflationary approaches" to the cognitive organization of scientific research, like those of Rorty and Rouse, which fail to defeat a circularity argument. The way they challenge the view of science's cognitive specificity is based upon a *petitio principii*, since they tacitly admit that only by assuming an "immanent essence" of science one can hold that scientific research is predicated on a philosophically significant cognitive *differentia specifica*. Therefore, the starting point of such approaches does not differ from the starting point of those doctrines that defend the view of a "substantive rationality of science" in terms of normative epistemology and epistemological foundationalism. Consequently, the "deflationary approaches" go on to deconstruct the same "essence", which they by necessity assume to exist in order to formulate their tenets.

Just as a radical version of cognitive essentialism would be only that one which takes into consideration all five principal aspects, so a radical overcoming of cognitive essentialism would be that one which opposes all five aspects. Is such a radical overcoming feasible at all? And if yes, does the radical overcoming of cognitive essentialism allow a development of a consistent strategy for defending science's cognitive specificity?

6. APPROACHING THE CONCEPT OF COGNITIVE EXISTENTIALISM

In trying to answer these questions and to figure out a non-essentialist strategy, let me now take up the issue of science's cognitive specificity from a different perspective than the one adopted in the preceding two sections. *Vis-a-vis* the requirement to oppose all five principal aspects, this should be a perspective that will pay more attention to the fact that cognitive essentialism (like any other form of essentialism) is opposed to existentialism, i.e., the view that there is no essence that precedes existence. My first aim in this section is to define a general notion of (methodological) existentialism. That job accomplished, I will suggest a concept of cognitive existentialism in whose framework science's cognitive specificity can be defended.

By "existentialism" I mean, in the present context, a "radically antiessentialist" viewpoint whose main characteristic is the search for the "existential genesis" of every supposed "essence" in human life. As a radical alternative to each kind of (historical, cognitive, moral, aesthetic, political or theological) essentialism, "existentialism" is not to be confused with relativism, constructivism, pragmatism, conventionalist antirealism or any other non-radical antiessentialist point of view. Seen from an existentialist perspective, science is construed as a mode of existence. Like any other mode of existence science is not "outside" the finitude of human being. From an existentialist point of view, science's cognitive specificity should be sought in the specific manner of constituting "worlds" within the plurality of practices involved in scientific research. Here "world" is to be understood as an ingredient of what Heidegger calls being-in-the-world, i.e., a mode of existence within a horizon of discursive practices leading to the constitution of a specific cultural reality ("world"). According to the preceding considerations, a "world" of scientific research (as a peculiar mode of existence) includes scientific domains consituted by the same type of objectifying thematization.

Cognitive existentialism is a conception that treats scientific research as a mode of being-in-the-world-by-doing-research, which is neither reducible to a cognitive dynamics whose reification is an interconnection of true propositions, nor to an assemblage of "contingent factors", informing the life of scientific communities in different cultural, political, and administrative settings. Furthermore, cognitive existentialism places special emphasis upon the irreducibility of *constitution* (of scientific domains within the being-in-the-world-by-doing-research) to the "rational reconstruction" of science's cognitive dynamics or to empirical (psychologcal, sociological, cultural) science studies. The constitution of scientific domains always takes place within discursive-practical horizons of doing research. Each particular practice (designing differential equations, performing experiments and so on) brings into being new "meaningful facts" (for instance, facts about the states of a mechanical system) within the horizons of doing research. The constitution takes the form of an iterative hermeneutic circle: there is a circularity between carrying out a particular practice of research and the horizon of interrelated practices projected onto a scientific domain as a fore-structure of its further articulation. The very circularity is an ever-enriching process with an "anticipation of completeness" (Gadamer) that contributes to specifying (through the ongoing articulation of scientific domains) the mode of being-in-the-world. Heidegger (1962, 195) claims that what is decisive in the existential analysis of the modes of being-in-the-world is not to get out of the hermeneutic circle "but to come into it in the right way". Following this claim, I will try to show that to defend science's cognitive specificity in the framework of cognitive existentialism amounts to figuring out a way of entering into the hermeneutic circle of the constitution of scientific domains through objectifying thematization. A prerequisite for coming into this circle in the right way consists of giving an account of it without falling into vicious circularity. Now, the methodology of cognitive existentialism requires (what in phenomenological philosophy is called) a *paradigm of constitutional analysis.* In the next section, I will try to defend science's cognitive specificity by making use of the paradigm put forward by hermeneutic phenomenology.

7. ON THE HERMENEUTIC FORE-STRUCTURING OF SCIENCE'S COGNITIVE SPECIFICITY

The core of each phenomenological philosophy is a given paradigm of constitutional analysis, i.e., analysis of the constitution of meaning and meaningful objects. In Husserl's phenomenology, this paradigm is built upon a kind of transcendental idealism. It is an idealism that is concerned with the intentionality of consciousness' acts to the reality of the natural world. According transcendental-idealistic phenomenology, the meaningfulness of the natural world (the manifolds of meaningful objects) is constituted within the intentional life of consciousness. There is no meaning (however remote from the personal and social life of empirical individuals) in the world that is not derivable in a genetic manner from transcendental subjectivity. The locus of constitution is this subjectivity that endures as something self-identical. It is a "transcendental substance" that is a source of all intentionalities. According to the paradigm of constitutional analysis put forward by Husserl (Hua IV, 312), every manifold of objects has certain sorts of conscious acts correlated with it. "Phenomenology does not investigate the objects investigated by the researcher in other sciences; on the contrary, it investigates the total system of possible acts of consciousness, of possible appearances and meanings related to precisely those objects." Transcendental-idealistic phenomenology is a typical essentialist philosophical conception since it presupposes the independent existence of transcendental consciousness as an absolute essence.

An important consequence of Heidegger's reformulation of the transcendental phenomenology of consciousness' intentionality in a hermeneutic phenomenology of "facticity" is the rise of a new paradigm of constitutional analysis. The transcendental ego no longer plays the role of a privileged site of meaning constitution. It is rather the totality of existence (the "facticity") where this constitution takes place. In the new paradigm of constitutional analysis, there is a kind of interpretation that is an intrinsic moment of all human activities. This universal interpretative dimension of human being-in-the-world should not be confused with the concept of interpretation as a specific epistemic procedure. From the perspective of hermeneutic phenomenology, interpretation must be comprehended as a primordial existential act. Interpretation that is inherent in human being-in-the-world brings to light the meanings constituted by most elementary discursive practices. There is no meaningful practice within-the-world without interpretation. To make sense of this claim, one has to take into consideration Heidegger's account of the nexus "understanding-interpretation". Understanding is the projection of a horizon of possibilities of constituting meaning upon a given mode of being-in-the-world. By executing particular practices within this horizon, interpretation comes into play as the articulation of what is projected in understanding. Heidegger (1962, 188-89) observes: "In interpretation, understanding does not become something different. It becomes itself. Such interpretation is grounded existentially in understanding; the latter does not arise from the former. Nor is interpretation the acquiring of information about what is understood; it is rather the working-out of possibilities projected in understanding."

In rejecting the idea of an ultimate grounding provided by the ego's intentional life of consciousness, one focuses on the temporal-interpretative self-constitution of human being-in-the-world. In this perspective, the "existential structure of the primordial interpretation" (i.e., the structure revealed through *Daseinsanalytik*) involves three moments, which Heidegger calls fore-having (*Vorhabe*), fore-sight (*Vorsicht*), and fore-conception (*Vorgriff*). Fore-having is the background of familiar practices in which an "average understanding" of the situations of "everyday concernful dealing" is embedded; fore-sight provides the orientation of the everyday involvement in concernful practices; and fore-conception is the anticipatory grasp of what is supposed to be an outcome of the concernful practices. Thus considered, the existential structure of interpretation plays the role of *hermeneutic fore-structure* of meaning constitution taking place in the modes of being-in-the-world.

Heidegger's elaborations of the nexus "understanding-interpretation" provoke an interesting ambiguity in the way of construing the notion of hermeneutic fore-structure. On the one hand, this notion denotes the interrelatedness of the most elementary routine practices that form "everydayness" as the primordial mode of being-in-the-world. Speaking in terms of *Being and Time*, everydayness is the existential mode characterized by a concernful dealing with things that are "ready-to-hand". Within this mode, the pre-epistemic constitution of a meaningful (everyday) world comes into being. On this reading, everydayness is opposed to the scientific constituition of nature as an "epistemologically specified mode of existence". In transforming the ready-to-hand (and the meaningful world of everyday routine practices) into a thematically objectified presence-at-hand, scientific research guided by objectifying thematization remains based upon the concernful constitution of meaning. Yet the hermeneutic fore-structure of this constitution becomes "concealed" by objectifying thematization. In this regard, the notion of hermeneutic fore-structure can be construed in terms of the contradistinction between the pre-epistemic being-in-the-world and the world(s) constituted by objectifying thematization.

On the other hand, scientific research is a being-in-the-world that is predicated on its own discursive-practical everydayness. A regime of "everyday concernful dealing" (illustrated typically by a scientific community's everyday life) is to be attributed to scientific research as a mode of existence. In this concernful dealing (as interrelatedness of discursive practices), objectifying thematization takes place. Thus, the search for a hermeneutic fore-structure (within the everydayness) of scientific research is a completely reasonable task. This is why many important distinctions of *Daseinsanalytik* are to be applied not only to the inquiry into the "existentially primordial" mode of pre-thematizing (pre-objectifying, pre-epistemic) dealing with things that are ready-to-hand, but to the intrinsic discursive-practical fore-structuring of the cognitive structure of scientific research as well. In particular, the distinction between the "pre-predicative as-structure of seeing of the ready-to-hand" and the "thematic-predicative as-structure of seeing of the present-at-hand" is applicable to the existential-ontological interpretation of scientific research. Heidegger does not pay much attention to the "intrinsic everydayness" of science. Moreover, in his analysis of objectifying thematization the interrelatedness of discursive practices of the research process does not play any significant role. He focuses chiefly on the "existential genesis" of science (as "theoretical attitude" and "mathematical projection of nature") from the everyday concernful constitution of elementary meanings. (This is the guiding idea of Heidegger's "existential conception of science" in *Being and Time*.) Yet following the line of my preceding considerations, scientific research is characterized by its own everydayness, which exhibits important similarities to what Kuhn calls "normal science". Related to this "secondary everydayness", hermeneutic fore-structure is the horizon of possibilities that can be worked out by the objectifying thematization of scientific research. The working-out of these possibilities constitutes a domain of scientific research.

Now I am in a position to shed more light on the question of how the scrutiny of the hermeneutic fore-structure of scientific research helps one to devise an antiessentialist strategy for defending science's cognitive specificity. In looking at the everydayness of scientific research, one can recognize within the scientific community's anticipations, inclinations, and orientations the moments of fore-having, fore-sight, and fore-conception. Yet one is unable to isolate and represent these moments as a static structure *per se*. They are recognizable only in the processuality of scientific research. By the same token, the hermeneutic fore-structure is not "before" the domain's cognitive structure. It is rather the practical processuality of "structuring the cognitive structure". By carrying out a particular practice of scientific research, a particular possibility of the domain's articulation becomes actualized. Yet this does not imply a gradual "exhaustion" of the horizon of possibilities projected in the hermeneutic fore-structure. On the contrary, the more possibilities are actualized the wider becomes the horizon of possibilities for whose actualization a fore-having, a fore-sight, and a fore-concepion exist. This means that hermeneutic fore-structure and cognitive structure are involved in an ongoing interplay in the process of articulation of a scientific domain. In this interplay, there is, on the one hand, a constant transition from fore-having, fore-sight, and fore-conception of scientific research to the outcome of objectifying thematization, and a reverse transition from objectified knowledge (actualized possibilities) to activating new possibilities projected in the practical horizon of scientific research, on the other. The very articulation of a scientific domain can only be grasped by reflecting upon the interplay. At the bottom of it is the iterative hermeneutic circularity I indicated in the former section. This view has important consequences for scrutinizing objectifying thematization in terms of constitutional analysis.

In Section 2, I raised the claim that seen from an epistemological point of view, the four moments of objectifying thematization are to be cast in terms of a semantic view of theorizing. From that point of view, objectifying thematization is to be described as an established cognitive structure. Let me now shift the perspective from epistemology to hermeneutic phenomenology. In so doing, objectifying thematization begins to emerge as something that is both the process and the result of the interplay of hermeneutic fore-structure and cognitive structure. In the paradigm of constitutional analysis outlined thus far, objectifying thematization is a *thematizing project,* guiding the ongoing transformation of fore-having, fore-sight, and fore-conception of scientific research into cognitively structured domains (characterized by the four moments of objectifying thematization - principles of invariance, covariant differential equations, theoretical models, and explanatory scenarios). By means of such a project, the working-out of possibilities projected in hermeneutic fore-structure comes into play. The thematizing project is not an independent cognitive essence. It exists only within the interplay of hermeneutic fore-structure and cognitive structure.[12]. The thematizing project is the locus of the circular relations between the fore-structuring of scientific research as a mode of existence and the empirically manifested cognitive specificity of science. In analyzing this project in an interpretative manner, cognitive existentialism provides a methodological alternative to the strategies that seek to "explain" (i.e., to derive from principles) a science/nonscience demarcational line.

Let me return to the fact that in denying science's cognitive specificity, the non-radical versions of antiessentialism fail to defeat the circular argument (in the form of *petitio principii*). One has to admit, however, that to a certain extent circularity is unavoidable in any strategy of defending science's cognitive specificity. (I already announced the four moments of objectifying thematization as a hallmark of science.) Hence, as I pointed out, the task is not to eliminate circularity radically, but rather to reformulate it (by avoiding the logical paradox) as a hermeneutic circle that can come to grips with the intrinsic hermeneutic circularity of scientific research. This claim contains the idea of "double hermeneutics" - that is, the idea that one has to apply second level interpretative concepts (based upon a methodological implementation of a hermeneutic circle) to the first level of interpretative constitution (based upon an intrinsic circularity) of cognitive constructs. There is a "transcendental dimension" in this double hermeneutics.

In Jeff Malpas' view, this dimension refers to the circular relations between the conditions of having a specific form of experience (and mode of existence) and the empirically manifested specificity of it. The main point of this view is that the circular relations are not "behind" but "within" the empirical totality of the respective form of experience (and mode of existence). By implication, the transcendental dimension of double-hermeneutic existential analysis lies in the recognition that there are features of the cognitive body of science which render their possibility within the processual circularity inherent in scientific research. Since the circular relations are relations of "self-constitution" of a mode of existence, they are not determined by something outside them. Thus, there is a kind of transcendental argument involved in the double-hermeneutic investigation of the hermeneutic circularity through which a characteristic "world" comes into being. Actually, Malpas is interested not in the transcendental dimension of hermeneutic circularity but in the intrinsic interpretative nature of (what he calls) the "transcendental project" of reasoning. This is why he (1997, 17) writes that "only if we grasp the (transcendental) project as involving something like an 'interpretation' will it begin to make sense, for in interpretation it is precisely the integration of elements within a structure ... that is at issue." Paraphrasing Malpas, I would say that only if we grasp the analysis of the interpretative self-constitution of scientific research as involving something like a "transcendental reflection" (that does not reify a "transcendental substance") will it begin to make sense, for in this reflection it is the non-essentialist revealing of science's cognitive specificity that is at issue.

Given these considerations, let me take a closer look at how the transcendental dimension takes place in the existential analysis of science's cognitive specificity. Roughly speaking, the idea of "the transcendental" in this analysis consists of envisaging an investigation that proceeds from the fact of constituting a "world" within a thematizing project (or, the fact of actualizing possibilities projected in the hermeneutic fore-structure of scientific research) to the changing configurations of discursive practices in their processual interrelatedness. In this formulation, the thematizing project is both within the objectified "world" of scientific research (as specified moments of objectifying thematization) and within the interrelatedness of discursive practices. In other words, the thematizing project is within both the hermeneutic fore-structure and the objectified cognitive structure, but it can only be identified and analyzed within their interplay. In this regard, the transcendental dimension of cognitive existentialism consists of reflecting upon the aforementioned transitions that make possible the specificity of scientific research as a mode of existence. In reflecting upon these transitions, one aims not at defining a priori invariant conditions but at revealing a constituting process (the interplay of hermeneutic fore-structre and cognitive structure). In other words, if one goes on to demonstrate that "the transcendental" is not derivable from an "atemporal essence", but lies in the dynamics of discursive practices (characterized by a complex temporality) in which a cognitive structure gradually takes shape, then one would be able, from an existentialist point of view, to provide a defence of science's cognitive specificity. The transcendental reflection of cognitive existentialism does not make an appeal to a "deduction" of the thematizing project. This reflection only shows the possibility to articulate in an interpretative manner that hermeneutic circularity of scientific research in which a cognitive spicificity comes to the fore. The double-hermeneutic circularity of the transcendental reflection is not to be disentangled from the circularity of the self-consituting being-in-the-world.

My aim in this essay was to indicate the direction of the answer to the question of how to be simultaneously an antiessentialist and a champion of science's cognitive specificity. The very answer would require an "existential analysis" of the thematizing project that reflects not only upon the constitution of scientific domains but also upon the hermeneutic fore-structuring of the very moments of objectifying thematization.

NOTES

1 On the concept of "objectifying thematization", see Kockelmans, Heidegger, *124-130; Idem,* Ideas, 110-114.

2 I am following here van Fraassen's (1989) approach ("the semantic view") to scientific theorizing. An essential advantage of this approach is that it wards off a kind of Platonist reification of scientific laws. Elsewhere (Ginev, *Critique*, pp. 25-30), I advocate the view of complementarity between the van Fraassen's cognitive picture of scientific theorizing and the picture of science's discursive practices depicted by hermeneutic phenomenology.

3 Here I am following Wigner's (1967) approach to the principles of invariance.

4 See, for example, Knorr-Cetina 1981, pp. 110-114.

5 On these objections see Ginev, *Scientific Progress*, and Ginev, *Methodological Principles*.

6 See especially Worrall, *Value*.

7 A profound analysis of this claim is suggested by Arthur Danto *Analytic Philosophy*.

8 See Joseph Rouse, *Engaging*, ch. 1.

9 On a taxonomy of the kinds of cognitive relativism see Howard Sankey, *Rationality*, pp. 42-51.

10 See Fine, *Game*, pp. 147-148.

11 Taking into account the subtitle of Joseph Rouse's *Knowledge* - "Toward a Political Philosophy of Science" - and the title of a chapter of his *Engaging* - "The Politics of Postmodern Philosophy of Science", one is to reach the conclusion that the way from a "political philosophy" to a "politics of the postmodern philosophy" is Rouse's antiessentialist way of dissolving philosophy of science in cultural studies. A particular, but very important, task of the strategy of non-essentialist defending science's cognitive specificity I am looking for, is to "save the autonomy of the philosophy of science". In this regard, my efforts go in the opposite direction to Rouse's elaborations.

12 See on this point, Ginev *Critique*, pp. 33-44.

REFERENCES

Arthur Danto. "Analytic Philosophy." *Social Research*, Vol. 47, No.4 (1980): 611-625.

Fine, Arthur. *The Shaky Game (Einstein and the Quantum Theory)*. Chicago and London: The University of Chicago Press, 1986.

Ginev, Dimitri. "Scientific Progress and the Hermeneutic Circle." *Studies in History and Philosophy of Science*, Vol. 19 (1988): 390-395.

Ginev, Dimitri. "Do We Need Fixed Methodological Principles?" *British Journal for the Philosophy of Science*, Vol. 44 (1993), pp. 329-334.

Ginev, Dimitiri. *Critique of Epistemological Reason.* Bulgarian Academic Monograps (5). Sofia: Academic Press, 2000.

Heidegger, Martin. *Being and Time*. (Translated by John Macquarrie and Edward Robinson). San Francisco: Harper, 1962.

Husserl, Edmund. (Hua IV). *Ideen zu einer reinen Phänomenologie und phänomenologischen Philosophie. Zweites Buch: Phänomenologische Untersuchungen zur Konstitution*. M. Biemel (Hg.). The Hague: Nijhoff, 1952.

Knorr-Cetina, Karin. *The Manufacture of Knowledge*. Oxford: Pergamon, 1981.

Kockelmans, Joseph. *Heidegger and Science*. Washington, D.C.: University Press of America, 1985.

Kockelmans, Joseph. *Ideas for a Hermeneutic Phenomenology of the Natural Sciences*. Dordrecht/Boston/London: Kluwer, 1993.

Latour, Bruno. *Science in Action*. Cambridge, Mass.: Harvard University Press, 1987.

Malpas, Jeff. "The Transcendental Circle." *Australasian Journal of Philosophy*, Vol. 75, No. 1 (1997): 1-20.

Rorty, Richard. "Is Natural Science a Natural Kind?" in: Ernan McMullin (ed.). *Construction and Constraint (The Shaping of Scientific Rationality)*. Notre Dame, Indiana: University of Notre Dame Press, 1988.

Rouse, Joseph. *Knowledge and Power*. Ithaca and London: Cornell University Press, 1987.

Rouse, Joseph. *Engaging Science*. Ithaca and London: Cornell University Press, 1996.

Sankey, Howard. *Rationality, Relativism and Incommensurability*. Aldershot: Ashgate, 1997.

van Fraassen, Bas. *Laws and Symmetry*. Oxford: Clarendon Press, 1989.

Vogel, Steven. "Science, Practice and Politics." *Social Epistemology*, vol. 5(1991): 267-292.

Wigner, Eugen. *Symmetries and Reflections*. Bloomington, Ind.: Indiana University Press, 1967.

Worrall, John. "The Value of a Fixed Methodology." *British Journal for the Philosophy of Science*, Vol. 39 (1988): 263-275.

NOTES ON CONTRIBUTORS

MICHAEL BUSHEV taught philosophy of non-classical physics at St. Kliment Ohridski University of Sofia, is now Senior Research Fellow at the Institute of Solid State Physics, Bulgarian Academy of Sciences. Among his fields of interest are equilibrium and non-equilibrium phaze transitions, and the methodological problems of the application of mathematics in physics. He is author of *Synergetics: Chaos, Order, Self-Organization*, Singapore 1994.

VIHREN BOUZOV is Associate Professor at the Sts. Cyril and Methodius University of Veliko Turnovo, teaching logic and general philosophy of science. He is author of *Philosophy of Language*, Sofia 2002, *Aspects of Juridical Logic*, Sofia 1999; *Practical Logic*, Sofia 1998.

DAFINA GENOVA is Associate Professor of English and Linguistics at the Department of English at Sts. Cyril and Methodius University of Veliko Turnovo. Her main inteests are in spoken discourse and, more specifically, in speech act theory, pragmatics and conversation analysis as approaches to discourse. She is also interested in semantics of natural language and philosophy of language. She is author of *Meaning Representation and Use of Language*, Veliko Turnovo 1997.

DIMITRI GINEV has lived for many years in Bulgaria. He is the founding editor of the journals *Studia Culturologica* and *Divinatio*. He tries to combine in his research work (and his teaching activities at the St. Kliment Ohridski University of Sofia) studies in the history of hermeneutics with issues of the "Continental philosophy of science". He has co-edited (with Robert Cohen) *Issues and Images in the Philosophy of Science. Boston Studies in the Philosophy of Science*, vol. 192, 1997.

LILIA GUROVA is Associate Professor at the Department of Cognitive Science of the New Bulgarian University. She has published extensively on philosophy of technology, and methodology of cognitive science, including one monograph on *The Role of the Problems*, Sofia 1998.

ASSEN PETROV is Emeritus Professor at the Institute for Nuclear Research, Bulgarian Academy of Sciences. He has written on quantum logic and the mathematical foundations of quantum mechanics. Co-author (with Sava Petrov) of *Quantum Mechanics (Interpretations and Alternatives)*, Sofia 1989.

SAVA PETROV (1934 - 1989) was Professor of Philosophy of Science at the Bulgarian Academy of Sciences. He has written extensively on classical and non-

Dimitri Ginev (ed.), Bulgarian Studies in the Philosophy of Science, 207-209.

classical epistemology, the logical paradoxes, paraconsistent logic, Kant's theoretical philosophy and its impact on the contemporaray philosophy of science, the hidden ontologies of various methodological research programs, and the philosophical dimensions of the historiography of science. His books in philosophy of science include *The Logical Paradoxes from a Philosophical Point of View*, *Methodology of the Substrative Approach*, and (with Assen Petrov), *Quantum Mechanics (Interpretations and Alternatives)*.

AZARYA POLIKAROV (1921 - 2000) was a prominent figure at international philosophical meetings for more than fifty years. He was an actual member of several international philosophical institutions, including the Division of Philosophy at UNESCO, and the International Institute of Philosophy. He was Visiting Professor at the University of Leipzig and Humboldt-University of Berlin. For a long period, his position in Bulgaria was a Director of the Center for Scientific Information at the Bulgarian Academy of Sciences. His books, published in six languages, include *Relativity and Quanta*, 1961, *Moderne Phsik - Weltbild und Denkstil*, 1966, *Essays on the Methodology of Science*, 1981, *Orientation in the Methodology of Science*, 1987, *Scientific Revioltions and Progress*, 1993, *Einsteins Theorien und Ansichten*, 2000. (On Polikarov's position of "heuristic methodology of science", and his defence of scientific, philosophical, and political pluralism see Ginev's paper in *Journal for General Philosophy of Science*, XVII(1986).

SLAVIAN RADEV is a Senior Research Fellow at the Institute of Mathematics, Bulgarian Academy of Sciences. He teaches informatics and logic at the Institute of Computer Science of the Technical University of Bialystok. He is author of the *Foundations of Informatics*, Warsaw 2000.

VLADIMIR SOTIROV received his PhD in mathematical logic from the Lomonosov State University of Moscow, taught logic and philosophy of mathematics at the St Kliment Ohridski University of Sofia, and is now Associate Professor at the Department of Mathematical Logic of the Institute of Mathematics and Informatics, Bulgarian Academy of Sciences. Among his fields of interest are modal logics on intutionist base, Aristotelian syllogistic and voting systems. He also engages in research on the history of mathematics.

SPAS SPASSOV is a Research Fellow at the Bulgarian Academy of Sciences. He teaches now at the Universty of Montreal. His interests include the French Neorationalism (especially, Bachelard's philosophy), philosophy of biology (with special emphasis upon Jacob's ideas), and philosophy of technology.

ANGUEL STEFANOV is Head of the Department of Philosophy, Logic and Methodology of Science of the Institute for Philosophical Research, Bulgarian Academy of Sciences. He is also Professor at the Sts. Cyril and Methodius University of Veliko Turnovo, teaching Epistemology and Philosophy of Physics, and Vice-President of the Bulgarian Philosophical Association. Among the other books, he is the author of *The Challenge of Incommensurability*, Sofia 2000; *Faculty of Theorizing*, Sofia 1999; *What is Physical Reality?*, Gabrovo 1999; *Complementarity: Pro and Con*, Sofia 1998.

MARTIN TABAKOV is Senior Research Fellow at the Department of Philosophical Logic of the Bulgarian Academy of Sciences. He has published on paraconsistent and other non-classical logical systems, Goedel's theorem, the existence in mathematics, and on various issues in the history of logic. His most recent book is *Revolutions in Modern Logic*, 2002.

Boston Studies in the Philosophy of Science

Editor: Robert S. Cohen, *Boston University*

1. M.W. Wartofsky (ed.): *Proceedings of the Boston Colloquium for the Philosophy of Science, 1961/1962.* [Synthese Library 6] 1963 ISBN 90-277-0021-4
2. R.S. Cohen and M.W. Wartofsky (eds.): *Proceedings of the Boston Colloquium for the Philosophy of Science, 1962/1964.* In Honor of P. Frank. [Synthese Library 10] 1965 ISBN 90-277-9004-0
3. R.S. Cohen and M.W. Wartofsky (eds.): *Proceedings of the Boston Colloquium for the Philosophy of Science, 1964/1966.* In Memory of Norwood Russell Hanson. [Synthese Library 14] 1967 ISBN 90-277-0013-3
4. R.S. Cohen and M.W. Wartofsky (eds.): *Proceedings of the Boston Colloquium for the Philosophy of Science, 1966/1968.* [Synthese Library 18] 1969 ISBN 90-277-0014-1
5. R.S. Cohen and M.W. Wartofsky (eds.): *Proceedings of the Boston Colloquium for the Philosophy of Science, 1966/1968.* [Synthese Library 19] 1969 ISBN 90-277-0015-X
6. R.S. Cohen and R.J. Seeger (eds.): *Ernst Mach, Physicist and Philosopher.* [Synthese Library 27] 1970 ISBN 90-277-0016-8
7. M. Čapek: *Bergson and Modern Physics.* A Reinterpretation and Re-evaluation. [Synthese Library 37] 1971 ISBN 90-277-0186-5
8. R.C. Buck and R.S. Cohen (eds.): *PSA 1970.* Proceedings of the 2nd Biennial Meeting of the Philosophy and Science Association (Boston, Fall 1970). In Memory of Rudolf Carnap. [Synthese Library 39] 1971 ISBN 90-277-0187-3; Pb 90-277-0309-4
9. A.A. Zinov'ev: *Foundations of the Logical Theory of Scientific Knowledge (Complex Logic).* Translated from Russian. Revised and enlarged English Edition, with an Appendix by G.A. Smirnov, E.A. Sidorenko, A.M. Fedina and L.A. Bobrova. [Synthese Library 46] 1973 ISBN 90-277-0193-8; Pb 90-277-0324-8
10. L. Tondl: *Scientific Procedures.* A Contribution Concerning the Methodological Problems of Scientific Concepts and Scientific Explanation.Translated from Czech. [Synthese Library 47] 1973 ISBN 90-277-0147-4; Pb 90-277-0323-X
11. R.J. Seeger and R.S. Cohen (eds.): *Philosophical Foundations of Science.* Proceedings of Section L, 1969, American Association for the Advancement of Science. [Synthese Library 58] 1974 ISBN 90-277-0390-6; Pb 90-277-0376-0
12. A. Grünbaum: *Philosophical Problems of Space and Times.* 2nd enlarged ed. [Synthese Library 55] 1973 ISBN 90-277-0357-4; Pb 90-277-0358-2
13. R.S. Cohen and M.W. Wartofsky (eds.): *Logical and Epistemological Studies in Contemporary Physics.* Proceedings of the Boston Colloquium for the Philosophy of Science, 1969/72, Part I. [Synthese Library 59] 1974 ISBN 90-277-0391-4; Pb 90-277-0377-9
14. R.S. Cohen and M.W. Wartofsky (eds.): *Methodological and Historical Essays in the Natural and Social Sciences.* Proceedings of the Boston Colloquium for the Philosophy of Science, 1969/72, Part II. [Synthese Library 60] 1974 ISBN 90-277-0392-2; Pb 90-277-0378-7
15. R.S. Cohen, J.J. Stachel and M.W. Wartofsky (eds.): *For Dirk Struik.* Scientific, Historical and Political Essays in Honor of Dirk J. Struik. [Synthese Library 61] 1974 ISBN 90-277-0393-0; Pb 90-277-0379-5
16. N. Geschwind: *Selected Papers on Language and the Brains.* [Synthese Library 68] 1974 ISBN 90-277-0262-4; Pb 90-277-0263-2
17. B.G. Kuznetsov: *Reason and Being.* Translated from Russian. Edited by C.R. Fawcett and R.S. Cohen. 1987 ISBN 90-277-2181-5

18. P. Mittelstaedt: *Philosophical Problems of Modern Physics.* Translated from the revised 4th German edition by W. Riemer and edited by R.S. Cohen. [Synthese Library 95] 1976
ISBN 90-277-0285-3; Pb 90-277-0506-2
19. H. Mehlberg: *Time, Causality, and the Quantum Theory.* Studies in the Philosophy of Science. Vol. I: *Essay on the Causal Theory of Time.* Vol. II: *Time in a Quantized Universe.* Translated from French. Edited by R.S. Cohen. 1980 Vol. I: ISBN 90-277-0721-9; Pb 90-277-1074-0
Vol. II: ISBN 90-277-1075-9; Pb 90-277-1076-7
20. K.F. Schaffner and R.S. Cohen (eds.): *PSA 1972.* Proceedings of the 3rd Biennial Meeting of the Philosophy of Science Association (Lansing, Michigan, Fall 1972). [Synthese Library 64] 1974 ISBN 90-277-0408-2; Pb 90-277-0409-0
21. R.S. Cohen and J.J. Stachel (eds.): *Selected Papers of Léon Rosenfeld.* [Synthese Library 100] 1979 ISBN 90-277-0651-4; Pb 90-277-0652-2
22. M. Čapek (ed.): *The Concepts of Space and Time.* Their Structure and Their Development. [Synthese Library 74] 1976 ISBN 90-277-0355-8; Pb 90-277-0375-2
23. M. Grene: *The Understanding of Nature.* Essays in the Philosophy of Biology. [Synthese Library 66] 1974 ISBN 90-277-0462-7; Pb 90-277-0463-5
24. D. Ihde: *Technics and Praxis.* A Philosophy of Technology. [Synthese Library 130] 1979
ISBN 90-277-0953-X; Pb 90-277-0954-8
25. J. Hintikka and U. Remes: *The Method of Analysis.* Its Geometrical Origin and Its General Significance. [Synthese Library 75] 1974 ISBN 90-277-0532-1; Pb 90-277-0543-7
26. J.E. Murdoch and E.D. Sylla (eds.): *The Cultural Context of Medieval Learning.* Proceedings of the First International Colloquium on Philosophy, Science, and Theology in the Middle Ages, 1973. [Synthese Library 76] 1975 ISBN 90-277-0560-7; Pb 90-277-0587-9
27. M. Grene and E. Mendelsohn (eds.): *Topics in the Philosophy of Biology.* [Synthese Library 84] 1976 ISBN 90-277-0595-X; Pb 90-277-0596-8
28. J. Agassi: *Science in Flux.* [Synthese Library 80] 1975
ISBN 90-277-0584-4; Pb 90-277-0612-3
29. J.J. Wiatr (ed.): *Polish Essays in the Methodology of the Social Sciences.* [Synthese Library 131] 1979 ISBN 90-277-0723-5; Pb 90-277-0956-4
30. P. Janich: *Protophysics of Time.* Constructive Foundation and History of Time Measurement. Translated from German. 1985 ISBN 90-277-0724-3
31. R.S. Cohen and M.W. Wartofsky (eds.): *Language, Logic, and Method.* 1983
ISBN 90-277-0725-1
32. R.S. Cohen, C.A. Hooker, A.C. Michalos and J.W. van Evra (eds.): *PSA 1974.* Proceedings of the 4th Biennial Meeting of the Philosophy of Science Association. [Synthese Library 101] 1976 ISBN 90-277-0647-6; Pb 90-277-0648-4
33. G. Holton and W.A. Blanpied (eds.): *Science and Its Public.* The Changing Relationship. [Synthese Library 96] 1976 ISBN 90-277-0657-3; Pb 90-277-0658-1
34. M.D. Grmek, R.S. Cohen and G. Cimino (eds.): *On Scientific Discovery.* The 1977 Erice Lectures. 1981 ISBN 90-277-1122-4; Pb 90-277-1123-2
35. S. Amsterdamski: *Between Experience and Metaphysics.* Philosophical Problems of the Evolution of Science. Translated from Polish. [Synthese Library 77] 1975
ISBN 90-277-0568-2; Pb 90-277-0580-1
36. M. Marković and G. Petrović (eds.): *Praxis.* Yugoslav Essays in the Philosophy and Methodology of the Social Sciences. [Synthese Library 134] 1979
ISBN 90-277-0727-8; Pb 90-277-0968-8

Boston Studies in the Philosophy of Science

37. H. von Helmholtz: *Epistemological Writings.* The Paul Hertz / Moritz Schlick Centenary Edition of 1921. Translated from German by M.F. Lowe. Edited with an Introduction and Bibliography by R.S. Cohen and Y. Elkana. [Synthese Library 79] 1977
ISBN 90-277-0290-X; Pb 90-277-0582-8
38. R.M. Martin: *Pragmatics, Truth and Language.* 1979
ISBN 90-277-0992-0; Pb 90-277-0993-9
39. R.S. Cohen, P.K. Feyerabend and M.W. Wartofsky (eds.): *Essays in Memory of Imre Lakatos.* [Synthese Library 99] 1976 ISBN 90-277-0654-9; Pb 90-277-0655-7
40. Not published.
41. Not published.
42. H.R. Maturana and F.J. Varela: *Autopoiesis and Cognition.* The Realization of the Living. With a Preface to "Autopoiesis' by S. Beer. 1980 ISBN 90-277-1015-5; Pb 90-277-1016-3
43. A. Kasher (ed.): *Language in Focus: Foundations, Methods and Systems.* Essays in Memory of Yehoshua Bar-Hillel. [Synthese Library 89] 1976
ISBN 90-277-0644-1; Pb 90-277-0645-X
44. T.D. Thao: *Investigations into the Origin of Language and Consciousness.* 1984
ISBN 90-277-0827-4
45. F.G.-I. Nagasaka (ed.): *Japanese Studies in the Philosophy of Science.* 1997
ISBN 0-7923-4781-1
46. P.L. Kapitza: *Experiment, Theory, Practice.* Articles and Addresses. Edited by R.S. Cohen. 1980 ISBN 90-277-1061-9; Pb 90-277-1062-7
47. M.L. Dalla Chiara (ed.): *Italian Studies in the Philosophy of Science.* 1981
ISBN 90-277-0735-9; Pb 90-277-1073-2
48. M.W. Wartofsky: *Models.* Representation and the Scientific Understanding. [Synthese Library 129] 1979 ISBN 90-277-0736-7; Pb 90-277-0947-5
49. T.D. Thao: *Phenomenology and Dialectical Materialism.* Edited by R.S. Cohen. 1986
ISBN 90-277-0737-5
50. Y. Fried and J. Agassi: *Paranoia.* A Study in Diagnosis. [Synthese Library 102] 1976
ISBN 90-277-0704-9; Pb 90-277-0705-7
51. K.H. Wolff: *Surrender and Cath.* Experience and Inquiry Today. [Synthese Library 105] 1976
ISBN 90-277-0758-8; Pb 90-277-0765-0
52. K. Kosík: *Dialectics of the Concrete.* A Study on Problems of Man and World. 1976
ISBN 90-277-0761-8; Pb 90-277-0764-2
53. N. Goodman: *The Structure of Appearance.* [Synthese Library 107] 1977
ISBN 90-277-0773-1; Pb 90-277-0774-X
54. H.A. Simon: *Models of Discovery* and Other Topics in the Methods of Science. [Synthese Library 114] 1977 ISBN 90-277-0812-6; Pb 90-277-0858-4
55. M. Lazerowitz: *The Language of Philosophy.* Freud and Wittgenstein. [Synthese Library 117] 1977 ISBN 90-277-0826-6; Pb 90-277-0862-2
56. T. Nickles (ed.): *Scientific Discovery, Logic, and Rationality.* 1980
ISBN 90-277-1069-4; Pb 90-277-1070-8
57. J. Margolis: *Persons and Mind.* The Prospects of Nonreductive Materialism. [Synthese Library 121] 1978 ISBN 90-277-0854-1; Pb 90-277-0863-0
58. G. Radnitzky and G. Andersson (eds.): *Progress and Rationality in Science.* [Synthese Library 125] 1978 ISBN 90-277-0921-1; Pb 90-277-0922-X
59. G. Radnitzky and G. Andersson (eds.): *The Structure and Development of Science.* [Synthese Library 136] 1979 ISBN 90-277-0994-7; Pb 90-277-0995-5

Boston Studies in the Philosophy of Science

60. T. Nickles (ed.): *Scientific Discovery.* Case Studies. 1980
ISBN 90-277-1092-9; Pb 90-277-1093-7
61. M.A. Finocchiaro: *Galileo and the Art of Reasoning.* Rhetorical Foundation of Logic and Scientific Method. 1980 ISBN 90-277-1094-5; Pb 90-277-1095-3
62. W.A. Wallace: *Prelude to Galileo.* Essays on Medieval and 16th-Century Sources of Galileo's Thought. 1981 ISBN 90-277-1215-8; Pb 90-277-1216-6
63. F. Rapp: *Analytical Philosophy of Technology.* Translated from German. 1981
ISBN 90-277-1221-2; Pb 90-277-1222-0
64. R.S. Cohen and M.W. Wartofsky (eds.): *Hegel and the Sciences.* 1984 ISBN 90-277-0726-X
65. J. Agassi: *Science and Society.* Studies in the Sociology of Science. 1981
ISBN 90-277-1244-1; Pb 90-277-1245-X
66. L. Tondl: *Problems of Semantics.* A Contribution to the Analysis of the Language of Science. Translated from Czech. 1981 ISBN 90-277-0148-2; Pb 90-277-0316-7
67. J. Agassi and R.S. Cohen (eds.): *Scientific Philosophy Today.* Essays in Honor of Mario Bunge. 1982 ISBN 90-277-1262-X; Pb 90-277-1263-8
68. W. Krajewski (ed.): *Polish Essays in the Philosophy of the Natural Sciences.* Translated from Polish and edited by R.S. Cohen and C.R. Fawcett. 1982
ISBN 90-277-1286-7; Pb 90-277-1287-5
69. J.H. Fetzer: *Scientific Knowledge.* Causation, Explanation and Corroboration. 1981
ISBN 90-277-1335-9; Pb 90-277-1336-7
70. S. Grossberg: *Studies of Mind and Brain.* Neural Principles of Learning, Perception, Development, Cognition, and Motor Control. 1982 ISBN 90-277-1359-6; Pb 90-277-1360-X
71. R.S. Cohen and M.W. Wartofsky (eds.): *Epistemology, Methodology, and the Social Sciences.* 1983. ISBN 90-277-1454-1
72. K. Berka: *Measurement.* Its Concepts, Theories and Problems. Translated from Czech. 1983
ISBN 90-277-1416-9
73. G.L. Pandit: *The Structure and Growth of Scientific Knowledge.* A Study in the Methodology of Epistemic Appraisal. 1983 ISBN 90-277-1434-7
74. A.A. Zinov'ev: *Logical Physics.* Translated from Russian. Edited by R.S. Cohen. 1983 [*see also* Volume 9] ISBN 90-277-0734-0
75. G-G. Granger: *Formal Thought and the Sciences of Man.* Translated from French. With and Introduction by A. Rosenberg. 1983 ISBN 90-277-1524-6
76. R.S. Cohen and L. Laudan (eds.): *Physics, Philosophy and Psychoanalysis.* Essays in Honor of Adolf Grünbaum. 1983 ISBN 90-277-1533-5
77. G. Böhme, W. van den Daele, R. Hohlfeld, W. Krohn and W. Schäfer: *Finalization in Science.* The Social Orientation of Scientific Progress. Translated from German. Edited by W. Schäfer. 1983 ISBN 90-277-1549-1
78. D. Shapere: *Reason and the Search for Knowledge.* Investigations in the Philosophy of Science. 1984 ISBN 90-277-1551-3; Pb 90-277-1641-2
79. G. Andersson (ed.): *Rationality in Science and Politics.* Translated from German. 1984
ISBN 90-277-1575-0; Pb 90-277-1953-5
80. P.T. Durbin and F. Rapp (eds.): *Philosophy and Technology.* [*Also* Philosophy and Technology Series, Vol. 1] 1983 ISBN 90-277-1576-9
81. M. Marković: *Dialectical Theory of Meaning.* Translated from Serbo-Croat. 1984
ISBN 90-277-1596-3
82. R.S. Cohen and M.W. Wartofsky (eds.): *Physical Sciences and History of Physics.* 1984.
ISBN 90-277-1615-3

Boston Studies in the Philosophy of Science

83. É. Meyerson: *The Relativistic Deduction.* Epistemological Implications of the Theory of Relativity. Translated from French. With a Review by Albert Einstein and an Introduction by Milić Čapek. 1985 ISBN 90-277-1699-4
84. R.S. Cohen and M.W. Wartofsky (eds.): *Methodology, Metaphysics and the History of Science.* In Memory of Benjamin Nelson. 1984 ISBN 90-277-1711-7
85. G. Tamás: *The Logic of Categories.* Translated from Hungarian. Edited by R.S. Cohen. 1986 ISBN 90-277-1742-7
86. S.L. de C. Fernandes: *Foundations of Objective Knowledge.* The Relations of Popper's Theory of Knowledge to That of Kant. 1985 ISBN 90-277-1809-1
87. R.S. Cohen and T. Schnelle (eds.): *Cognition and Fact.* Materials on Ludwik Fleck. 1986 ISBN 90-277-1902-0
88. G. Freudenthal: *Atom and Individual in the Age of Newton.* On the Genesis of the Mechanistic World View. Translated from German. 1986 ISBN 90-277-1905-5
89. A. Donagan, A.N. Perovich Jr and M.V. Wedin (eds.): *Human Nature and Natural Knowledge.* Essays presented to Marjorie Grene on the Occasion of Her 75th Birthday. 1986 ISBN 90-277-1974-8
90. C. Mitcham and A. Hunning (eds.): *Philosophy and Technology II.* Information Technology and Computers in Theory and Practice. [*Also* Philosophy and Technology Series, Vol. 2] 1986 ISBN 90-277-1975-6
91. M. Grene and D. Nails (eds.): *Spinoza and the Sciences.* 1986 ISBN 90-277-1976-4
92. S.P. Turner: *The Search for a Methodology of Social Science.* Durkheim, Weber, and the 19th-Century Problem of Cause, Probability, and Action. 1986. ISBN 90-277-2067-3
93. I.C. Jarvie: *Thinking about Society.* Theory and Practice. 1986 ISBN 90-277-2068-1
94. E. Ullmann-Margalit (ed.): *The Kaleidoscope of Science.* The Israel Colloquium: Studies in History, Philosophy, and Sociology of Science, Vol. 1. 1986 ISBN 90-277-2158-0; Pb 90-277-2159-9
95. E. Ullmann-Margalit (ed.): *The Prism of Science.* The Israel Colloquium: Studies in History, Philosophy, and Sociology of Science, Vol. 2. 1986 ISBN 90-277-2160-2; Pb 90-277-2161-0
96. G. Márkus: *Language and Production.* A Critique of the Paradigms. Translated from French. 1986 ISBN 90-277-2169-6
97. F. Amrine, F.J. Zucker and H. Wheeler (eds.): *Goethe and the Sciences: A Reappraisal.* 1987 ISBN 90-277-2265-X; Pb 90-277-2400-8
98. J.C. Pitt and M. Pera (eds.): *Rational Changes in Science.* Essays on Scientific Reasoning. Translated from Italian. 1987 ISBN 90-277-2417-2
99. O. Costa de Beauregard: *Time, the Physical Magnitude.* 1987 ISBN 90-277-2444-X
100. A. Shimony and D. Nails (eds.): *Naturalistic Epistemology.* A Symposium of Two Decades. 1987 ISBN 90-277-2337-0
101. N. Rotenstreich: *Time and Meaning in History.* 1987 ISBN 90-277-2467-9
102. D.B. Zilberman: *The Birth of Meaning in Hindu Thought.* Edited by R.S. Cohen. 1988 ISBN 90-277-2497-0
103. T.F. Glick (ed.): *The Comparative Reception of Relativity.* 1987 ISBN 90-277-2498-9
104. Z. Harris, M. Gottfried, T. Ryckman, P. Mattick Jr, A. Daladier, T.N. Harris and S. Harris: *The Form of Information in Science.* Analysis of an Immunology Sublanguage. With a Preface by Hilary Putnam. 1989 ISBN 90-277-2516-0
105. F. Burwick (ed.): *Approaches to Organic Form.* Permutations in Science and Culture. 1987 ISBN 90-277-2541-1

Boston Studies in the Philosophy of Science

106. M. Almási: *The Philosophy of Appearances.* Translated from Hungarian. 1989
ISBN 90-277-2150-5
107. S. Hook, W.L. O'Neill and R. O'Toole (eds.): *Philosophy, History and Social Action.* Essays in Honor of Lewis Feuer. With an Autobiographical Essay by L. Feuer. 1988
ISBN 90-277-2644-2
108. I. Hronszky, M. Fehér and B. Dajka: *Scientific Knowledge Socialized.* Selected Proceedings of the 5th Joint International Conference on the History and Philosophy of Science organized by the IUHPS (Veszprém, Hungary, 1984). 1988 ISBN 90-277-2284-6
109. P. Tillers and E.D. Green (eds.): *Probability and Inference in the Law of Evidence.* The Uses and Limits of Bayesianism. 1988 ISBN 90-277-2689-2
110. E. Ullmann-Margalit (ed.): *Science in Reflection.* The Israel Colloquium: Studies in History, Philosophy, and Sociology of Science, Vol. 3. 1988
ISBN 90-277-2712-0; Pb 90-277-2713-9
111. K. Gavroglu, Y. Goudaroulis and P. Nicolacopoulos (eds.): *Imre Lakatos and Theories of Scientific Change.* 1989 ISBN 90-277-2766-X
112. B. Glassner and J.D. Moreno (eds.): *The Qualitative-Quantitative Distinction in the Social Sciences.* 1989 ISBN 90-277-2829-1
113. K. Arens: *Structures of Knowing.* Psychologies of the 19th Century. 1989
ISBN 0-7923-0009-2
114. A. Janik: *Style, Politics and the Future of Philosophy.* 1989 ISBN 0-7923-0056-4
115. F. Amrine (ed.): *Literature and Science as Modes of Expression.* With an Introduction by S. Weininger. 1989 ISBN 0-7923-0133-1
116. J.R. Brown and J. Mittelstrass (eds.): *An Intimate Relation.* Studies in the History and Philosophy of Science. Presented to Robert E. Butts on His 60th Birthday. 1989
ISBN 0-7923-0169-2
117. F. D'Agostino and I.C. Jarvie (eds.): *Freedom and Rationality.* Essays in Honor of John Watkins. 1989 ISBN 0-7923-0264-8
118. D. Zolo: *Reflexive Epistemology.* The Philosophical Legacy of Otto Neurath. 1989
ISBN 0-7923-0320-2
119. M. Kearn, B.S. Philips and R.S. Cohen (eds.): *Georg Simmel and Contemporary Sociology.* 1989 ISBN 0-7923-0407-1
120. T.H. Levere and W.R. Shea (eds.): *Nature, Experiment and the Science.* Essays on Galileo and the Nature of Science. In Honour of Stillman Drake. 1989 ISBN 0-7923-0420-9
121. P. Nicolacopoulos (ed.): *Greek Studies in the Philosophy and History of Science.* 1990
ISBN 0-7923-0717-8
122. R. Cooke and D. Costantini (eds.): *Statistics in Science.* The Foundations of Statistical Methods in Biology, Physics and Economics. 1990 ISBN 0-7923-0797-6
123. P. Duhem: *The Origins of Statics.* Translated from French by G.F. Leneaux, V.N. Vagliente and G.H. Wagner. With an Introduction by S.L. Jaki. 1991 ISBN 0-7923-0898-0
124. H. Kamerlingh Onnes: *Through Measurement to Knowledge.* The Selected Papers, 1853-1926. Edited and with an Introduction by K. Gavroglu and Y. Goudaroulis. 1991
ISBN 0-7923-0825-5
125. M. Čapek: *The New Aspects of Time: Its Continuity and Novelties.* Selected Papers in the Philosophy of Science. 1991 ISBN 0-7923-0911-1
126. S. Unguru (ed.): *Physics, Cosmology and Astronomy, 1300–1700.* Tension and Accommodation. 1991 ISBN 0-7923-1022-5

Boston Studies in the Philosophy of Science

127. Z. Bechler: *Newton's Physics on the Conceptual Structure of the Scientific Revolution.* 1991 ISBN 0-7923-1054-3
128. É. Meyerson: *Explanation in the Sciences.* Translated from French by M-A. Siple and D.A. Siple. 1991 ISBN 0-7923-1129-9
129. A.I. Tauber (ed.): *Organism and the Origins of Self.* 1991 ISBN 0-7923-1185-X
130. F.J. Varela and J-P. Dupuy (eds.): *Understanding Origins.* Contemporary Views on the Origin of Life, Mind and Society. 1992 ISBN 0-7923-1251-1
131. G.L. Pandit: *Methodological Variance.* Essays in Epistemological Ontology and the Methodology of Science. 1991 ISBN 0-7923-1263-5
132. G. Munévar (ed.): *Beyond Reason.* Essays on the Philosophy of Paul Feyerabend. 1991 ISBN 0-7923-1272-4
133. T.E. Uebel (ed.): *Rediscovering the Forgotten Vienna Circle.* Austrian Studies on Otto Neurath and the Vienna Circle. Partly translated from German. 1991 ISBN 0-7923-1276-7
134. W.R. Woodward and R.S. Cohen (eds.): *World Views and Scientific Discipline Formation.* Science Studies in the [former] German Democratic Republic. Partly translated from German by W.R. Woodward. 1991 ISBN 0-7923-1286-4
135. P. Zambelli: *The Speculum Astronomiae and Its Enigma.* Astrology, Theology and Science in Albertus Magnus and His Contemporaries. 1992 ISBN 0-7923-1380-1
136. P. Petitjean, C. Jami and A.M. Moulin (eds.): *Science and Empires.* Historical Studies about Scientific Development and European Expansion. ISBN 0-7923-1518-9
137. W.A. Wallace: *Galileo's Logic of Discovery and Proof.* The Background, Content, and Use of His Appropriated Treatises on Aristotle's *Posterior Analytics.* 1992 ISBN 0-7923-1577-4
138. W.A. Wallace: *Galileo's Logical Treatises.* A Translation, with Notes and Commentary, of His Appropriated Latin Questions on Aristotle's *Posterior Analytics.* 1992 ISBN 0-7923-1578-2
    Set (137 + 138) ISBN 0-7923-1579-0
139. M.J. Nye, J.L. Richards and R.H. Stuewer (eds.): *The Invention of Physical Science.* Intersections of Mathematics, Theology and Natural Philosophy since the Seventeenth Century. Essays in Honor of Erwin N. Hiebert. 1992 ISBN 0-7923-1753-X
140. G. Corsi, M.L. dalla Chiara and G.C. Ghirardi (eds.): *Bridging the Gap: Philosophy, Mathematics and Physics.* Lectures on the Foundations of Science. 1992 ISBN 0-7923-1761-0
141. C.-H. Lin and D. Fu (eds.): *Philosophy and Conceptual History of Science in Taiwan.* 1992 ISBN 0-7923-1766-1
142. S. Sarkar (ed.): *The Founders of Evolutionary Genetics.* A Centenary Reappraisal. 1992 ISBN 0-7923-1777-7
143. J. Blackmore (ed.): *Ernst Mach – A Deeper Look.* Documents and New Perspectives. 1992 ISBN 0-7923-1853-6
144. P. Kroes and M. Bakker (eds.): *Technological Development and Science in the Industrial Age.* New Perspectives on the Science–Technology Relationship. 1992 ISBN 0-7923-1898-6
145. S. Amsterdamski: *Between History and Method.* Disputes about the Rationality of Science. 1992 ISBN 0-7923-1941-9
146. E. Ullmann-Margalit (ed.): *The Scientific Enterprise.* The Bar-Hillel Colloquium: Studies in History, Philosophy, and Sociology of Science, Volume 4. 1992 ISBN 0-7923-1992-3
147. L. Embree (ed.): *Metaarchaeology.* Reflections by Archaeologists and Philosophers. 1992 ISBN 0-7923-2023-9
148. S. French and H. Kamminga (eds.): *Correspondence, Invariance and Heuristics.* Essays in Honour of Heinz Post. 1993 ISBN 0-7923-2085-9
149. M. Bunzl: *The Context of Explanation.* 1993 ISBN 0-7923-2153-7

Boston Studies in the Philosophy of Science

150. I.B. Cohen (ed.): *The Natural Sciences and the Social Sciences.* Some Critical and Historical Perspectives. 1994 ISBN 0-7923-2223-1
151. K. Gavroglu, Y. Christianidis and E. Nicolaidis (eds.): *Trends in the Historiography of Science.* 1994 ISBN 0-7923-2255-X
152. S. Poggi and M. Bossi (eds.): *Romanticism in Science.* Science in Europe, 1790–1840. 1994 ISBN 0-7923-2336-X
153. J. Faye and H.J. Folse (eds.): *Niels Bohr and Contemporary Philosophy.* 1994 ISBN 0-7923-2378-5
154. C.C. Gould and R.S. Cohen (eds.): *Artifacts, Representations, and Social Practice.* Essays for Marx W. Wartofsky. 1994 ISBN 0-7923-2481-1
155. R.E. Butts: *Historical Pragmatics.* Philosophical Essays. 1993 ISBN 0-7923-2498-6
156. R. Rashed: *The Development of Arabic Mathematics: Between Arithmetic and Algebra.* Translated from French by A.F.W. Armstrong. 1994 ISBN 0-7923-2565-6
157. I. Szumilewicz-Lachman (ed.): *Zygmunt Zawirski: His Life and Work.* With Selected Writings on Time, Logic and the Methodology of Science. Translations by Feliks Lachman. Ed. by R.S. Cohen, with the assistance of B. Bergo. 1994 ISBN 0-7923-2566-4
158. S.N. Haq: *Names, Natures and Things.* The Alchemist Jābir ibn Ḥayyān and His *Kitāb al-Aḥjār* (Book of Stones). 1994 ISBN 0-7923-2587-7
159. P. Plaass: *Kant's Theory of Natural Science.* Translation, Analytic Introduction and Commentary by Alfred E. and Maria G. Miller. 1994 ISBN 0-7923-2750-0
160. J. Misiek (ed.): *The Problem of Rationality in Science and its Philosophy.* On Popper vs. Polanyi. The Polish Conferences 1988–89. 1995 ISBN 0-7923-2925-2
161. I.C. Jarvie and N. Laor (eds.): *Critical Rationalism, Metaphysics and Science.* Essays for Joseph Agassi, Volume I. 1995 ISBN 0-7923-2960-0
162. I.C. Jarvie and N. Laor (eds.): *Critical Rationalism, the Social Sciences and the Humanities.* Essays for Joseph Agassi, Volume II. 1995 ISBN 0-7923-2961-9
     Set (161–162) ISBN 0-7923-2962-7
163. K. Gavroglu, J. Stachel and M.W. Wartofsky (eds.): *Physics, Philosophy, and the Scientific Community.* Essays in the Philosophy and History of the Natural Sciences and Mathematics. In Honor of Robert S. Cohen. 1995 ISBN 0-7923-2988-0
164. K. Gavroglu, J. Stachel and M.W. Wartofsky (eds.): *Science, Politics and Social Practice.* Essays on Marxism and Science, Philosophy of Culture and the Social Sciences. In Honor of Robert S. Cohen. 1995 ISBN 0-7923-2989-9
165. K. Gavroglu, J. Stachel and M.W. Wartofsky (eds.): *Science, Mind and Art.* Essays on Science and the Humanistic Understanding in Art, Epistemology, Religion and Ethics. Essays in Honor of Robert S. Cohen. 1995 ISBN 0-7923-2990-2
     Set (163–165) ISBN 0-7923-2991-0
166. K.H. Wolff: *Transformation in the Writing.* A Case of Surrender-and-Catch. 1995 ISBN 0-7923-3178-8
167. A.J. Kox and D.M. Siegel (eds.): *No Truth Except in the Details.* Essays in Honor of Martin J. Klein. 1995 ISBN 0-7923-3195-8
168. J. Blackmore: *Ludwig Boltzmann, His Later Life and Philosophy, 1900–1906.* Book One: A Documentary History. 1995 ISBN 0-7923-3231-8
169. R.S. Cohen, R. Hilpinen and R. Qiu (eds.): *Realism and Anti-Realism in the Philosophy of Science.* Beijing International Conference, 1992. 1996 ISBN 0-7923-3233-4
170. I. Kuçuradi and R.S. Cohen (eds.): *The Concept of Knowledge.* The Ankara Seminar. 1995 ISBN 0-7923-3241-5

Boston Studies in the Philosophy of Science

171. M.A. Grodin (ed.): *Meta Medical Ethics*: The Philosophical Foundations of Bioethics. 1995 ISBN 0-7923-3344-6
172. S. Ramirez and R.S. Cohen (eds.): *Mexican Studies in the History and Philosophy of Science.* 1995 ISBN 0-7923-3462-0
173. C. Dilworth: *The Metaphysics of Science.* An Account of Modern Science in Terms of Principles, Laws and Theories. 1995 ISBN 0-7923-3693-3
174. J. Blackmore: *Ludwig Boltzmann, His Later Life and Philosophy, 1900–1906* Book Two: The Philosopher. 1995 ISBN 0-7923-3464-7
175. P. Damerow: *Abstraction and Representation.* Essays on the Cultural Evolution of Thinking. 1996 ISBN 0-7923-3816-2
176. M.S. Macrakis: *Scarcity's Ways: The Origins of Capital.* A Critical Essay on Thermodynamics, Statistical Mechanics and Economics. 1997 ISBN 0-7923-4760-9
177. M. Marion and R.S. Cohen (eds.): *Québec Studies in the Philosophy of Science.* Part I: Logic, Mathematics, Physics and History of Science. Essays in Honor of Hugues Leblanc. 1995 ISBN 0-7923-3559-7
178. M. Marion and R.S. Cohen (eds.): *Québec Studies in the Philosophy of Science.* Part II: Biology, Psychology, Cognitive Science and Economics. Essays in Honor of Hugues Leblanc. 1996 ISBN 0-7923-3560-0
Set (177–178) ISBN 0-7923-3561-9
179. Fan Dainian and R.S. Cohen (eds.): *Chinese Studies in the History and Philosophy of Science and Technology.* 1996 ISBN 0-7923-3463-9
180. P. Forman and J.M. Sánchez-Ron (eds.): *National Military Establishments and the Advancement of Science and Technology.* Studies in 20th Century History. 1996 ISBN 0-7923-3541-4
181. E.J. Post: *Quantum Reprogramming.* Ensembles and Single Systems: A Two-Tier Approach to Quantum Mechanics. 1995 ISBN 0-7923-3565-1
182. A.I. Tauber (ed.): *The Elusive Synthesis: Aesthetics and Science.* 1996 ISBN 0-7923-3904-5
183. S. Sarkar (ed.): *The Philosophy and History of Molecular Biology: New Perspectives.* 1996 ISBN 0-7923-3947-9
184. J.T. Cushing, A. Fine and S. Goldstein (eds.): *Bohmian Mechanics and Quantum Theory: An Appraisal.* 1996 ISBN 0-7923-4028-0
185. K. Michalski: *Logic and Time.* An Essay on Husserl's Theory of Meaning. 1996 ISBN 0-7923-4082-5
186. G. Munévar (ed.): *Spanish Studies in the Philosophy of Science.* 1996 ISBN 0-7923-4147-3
187. G. Schubring (ed.): *Hermann Günther Graßmann (1809–1877): Visionary Mathematician, Scientist and Neohumanist Scholar.* Papers from a Sesquicentennial Conference. 1996 ISBN 0-7923-4261-5
188. M. Bitbol: *Schrödinger's Philosophy of Quantum Mechanics.* 1996 ISBN 0-7923-4266-6
189. J. Faye, U. Scheffler and M. Urchs (eds.): *Perspectives on Time.* 1997 ISBN 0-7923-4330-1
190. K. Lehrer and J.C. Marek (eds.): *Austrian Philosophy Past and Present.* Essays in Honor of Rudolf Haller. 1996 ISBN 0-7923-4347-6
191. J.L. Lagrange: *Analytical Mechanics.* Translated and edited by Auguste Boissonade and Victor N. Vagliente. Translated from the *Mécanique Analytique, novelle édition* of 1811. 1997 ISBN 0-7923-4349-2
192. D. Ginev and R.S. Cohen (eds.): *Issues and Images in the Philosophy of Science.* Scientific and Philosophical Essays in Honour of Azarya Polikarov. 1997 ISBN 0-7923-4444-8

Boston Studies in the Philosophy of Science

193. R.S. Cohen, M. Horne and J. Stachel (eds.): *Experimental Metaphysics.* Quantum Mechanical Studies for Abner Shimony, Volume One. 1997 ISBN 0-7923-4452-9
194. R.S. Cohen, M. Horne and J. Stachel (eds.): *Potentiality, Entanglement and Passion-at-a-Distance.* Quantum Mechanical Studies for Abner Shimony, Volume Two. 1997 ISBN 0-7923-4453-7; Set 0-7923-4454-5
195. R.S. Cohen and A.I. Tauber (eds.): *Philosophies of Nature: The Human Dimension.* 1997 ISBN 0-7923-4579-7
196. M. Otte and M. Panza (eds.): *Analysis and Synthesis in Mathematics.* History and Philosophy. 1997 ISBN 0-7923-4570-3
197. A. Denkel: *The Natural Background of Meaning.* 1999 ISBN 0-7923-5331-5
198. D. Baird, R.I.G. Hughes and A. Nordmann (eds.): *Heinrich Hertz: Classical Physicist, Modern Philosopher.* 1999 ISBN 0-7923-4653-X
199. A. Franklin: *Can That be Right?* Essays on Experiment, Evidence, and Science. 1999 ISBN 0-7923-5464-8
200. D. Raven, W. Krohn and R.S. Cohen (eds.): *The Social Origins of Modern Science.* 2000 ISBN 0-7923-6457-0
201. Reserved
202. Reserved
203. B. Babich and R.S. Cohen (eds.): *Nietzsche, Theories of Knowledge, and Critical Theory.* Nietzsche and the Sciences I. 1999 ISBN 0-7923-5742-6
204. B. Babich and R.S. Cohen (eds.): *Nietzsche, Epistemology, and Philosophy of Science.* Nietzsche and the Science II. 1999 ISBN 0-7923-5743-4
205. R. Hooykaas: *Fact, Faith and Fiction in the Development of Science.* The Gifford Lectures given in the University of St Andrews 1976. 1999 ISBN 0-7923-5774-4
206. M. Fehér, O. Kiss and L. Ropolyi (eds.): *Hermeneutics and Science.* 1999 ISBN 0-7923-5798-1
207. R.M. MacLeod (ed.): *Science and the Pacific War.* Science and Survival in the Pacific, 1939-1945. 1999 ISBN 0-7923-5851-1
208. I. Hanzel: *The Concept of Scientific Law in the Philosophy of Science and Epistemology.* A Study of Theoretical Reason. 1999 ISBN 0-7923-5852-X
209. G. Helm; R.J. Deltete (ed./transl.): *The Historical Development of Energetics.* 1999 ISBN 0-7923-5874-0
210. A. Orenstein and P. Kotatko (eds.): *Knowledge, Language and Logic.* Questions for Quine. 1999 ISBN 0-7923-5986-0
211. R.S. Cohen and H. Levine (eds.): *Maimonides and the Sciences.* 2000 ISBN 0-7923-6053-2
212. H. Gourko, D.I. Williamson and A.I. Tauber (eds.): *The Evolutionary Biology Papers of Elie Metchnikoff.* 2000 ISBN 0-7923-6067-2
213. S. D'Agostino: *A History of the Ideas of Theoretical Physics.* Essays on the Nineteenth and Twentieth Century Physics. 2000 ISBN 0-7923-6094-X
214. S. Lelas: *Science and Modernity.* Toward An Integral Theory of Science. 2000 ISBN 0-7923-6303-5
215. E. Agazzi and M. Pauri (eds.): *The Reality of the Unobservable.* Observability, Unobservability and Their Impact on the Issue of Scientific Realism. 2000 ISBN 0-7923-6311-6
216. P. Hoyningen-Huene and H. Sankey (eds.): *Incommensurability and Related Matters.* 2001 ISBN 0-7923-6989-0
217. A. Nieto-Galan: *Colouring Textiles.* A History of Natural Dyestuffs in Industrial Europe. 2001 ISBN 0-7923-7022-8

Boston Studies in the Philosophy of Science

218. J. Blackmore, R. Itagaki and S. Tanaka (eds.): *Ernst Mach's Vienna 1895–1930*. Or Phenomenalism as Philosophy of Science. 2001 ISBN 0-7923-7122-4
219. R. Vihalemm (ed.): *Estonian Studies in the History and Philosophy of Science*. 2001 ISBN 0-7923-7189-5
220. W. Lefèvre (ed.): *Between Leibniz, Newton, and Kant*. Philosophy and Science in the Eighteenth Century. 2001 ISBN 0-7923-7198-4
221. T.F. Glick, M.Á. Puig-Samper and R. Ruiz (eds.): *The Reception of Darwinism in the Iberian World*. Spain, Spanish America and Brazil. 2001 ISBN 1-4020-0082-0
222. U. Klein (ed.): *Tools and Modes of Representation in the Laboratory Sciences*. 2001 ISBN 1-4020-0100-2
223. P. Duhem: *Mixture and Chemical Combination*. And Related Essays. Edited and translated, with an introduction, by Paul Needham. 2002 ISBN 1-4020-0232-7
224. J.C. Boudri: *What was Mechanical about Mechanics*. The Concept of Force Betweem Metaphysics and Mechanics from Newton to Lagrange. 2002 ISBN 1-4020-0233-5
225. B.E. Babich (ed.): *Hermeneutic Philosophy of Science, Van Gogh's Eyes, and God*. Essays in Honor of Patrick A. Heelan, S.J. 2002 ISBN 1-4020-0234-3
226. D. Davies Villemaire: *E.A. Burtt, Historian and Philosopher*. A Study of the Author of The Metaphysical Foundations of Modern Physical Science. 2002 ISBN 1-4020-0428-1
227. L.J. Cohen: *Knowledge and Language*. Selected Essays of L. Jonathan Cohen. Edited and with an introduction by James Logue. 2002 ISBN 1-4020-0474-5
228. G.E. Allen and R.M. MacLeod (eds.): *Science, History and Social Activism: A Tribute to Everett Mendelsohn*. 2002 ISBN 1-4020-0495-0
229. O. Gal: *Meanest Foundations and Nobler Superstructures*. Hooke, Newton and the "Compounding of the Celestiall Motions of the Planetts". 2002 ISBN 1-4020-0732-9
230. R. Nola: *Rescuing Reason*. A Critique of Anti-Rationalist Views of Science and Knowledge. 2003 Hb: ISBN 1-4020-1042-7; Pb ISBN 1-4020-1043-5
231. J. Agassi: *Science and Culture*. 2003 ISBN 1-4020-1156-3
232. M.C. Galavotti (ed.): *Observation and Experiment in the Natural and Social Science*. 2003 ISBN 1-4020-1251-9
233. A. Simões, A. Carneiro and M.P. Diogo (eds.): *Travels of Learning*. A Geography of Science in Europe. 2003 ISBN 1-4020-1259-4
234. A. Ashtekar, R. Cohen, D. Howard, J. Renn, S. Sarkar and A. Shimony (eds.): *Revisiting the Foundations of Relativistic Physics*. Festschrift in Honor of John Stachel. 2003 ISBN 1-4020-1284-5
235. R.P. Farell: *Feyerabend and Scientific Values*. Tightrope-Walking Rationality. 2003 ISBN 1-4020-1350-7
236. D. Ginev (ed.): *Bulgarian Studies in the Philosophy of Science*. 2003 ISBN 1-4020-1496-1

Also of interest:

R.S. Cohen and M.W. Wartofsky (eds.): *A Portrait of Twenty-Five Years Boston Colloquia for the Philosophy of Science, 1960-1985*. 1985 ISBN Pb 90-277-1971-3

Previous volumes are still available.

KLUWER ACADEMIC PUBLISHERS – DORDRECHT / BOSTON / LONDON

GPSR Compliance

The European Union's (EU) General Product Safety Regulation (GPSR) is a set of rules that requires consumer products to be safe and our obligations to ensure this.

If you have any concerns about our products, you can contact us on ProductSafety@springernature.com

In case Publisher is established outside the EU, the EU authorized representative is:

Springer Nature Customer Service Center GmbH
Europaplatz 3
69115 Heidelberg, Germany

Batch number: 10823959

Printed by Printforce, the Netherlands